Engineering and Operations of System of Systems

Engineering and Operations of System of Systems

By

John P.T. Mo

Professor of Manufacturing Engineering
RMIT University Australia

Ronald C. Beckett

Adjunct Professor
Swinburne University of Technology
Australia

CRC Press is an imprint of the
Taylor & Francis Group, an **informa** business

CRC Press
Taylor & Francis Group
6000 Broken Sound Parkway NW, Suite 300
Boca Raton, FL 33487-2742

First issued in paperback 2020

ISBN 13: 978-0-367-57086-6 (pbk)
ISBN 13: 978-1-138-63473-2 (hbk)

Library of Congress Cataloging-in-Publication Data

Names: Mo, John (Professor of Manufacturing Engineering), author.
Title: Engineering and operations of system of systems / John Mo and Ronald Beckett.
Description: Boca Raton : Taylor & Francis, 2019. | Includes bibliographical references and index.
Identifiers: LCCN 2018027840| ISBN 9781138634732 (hardback : alk. paper) | ISBN 9781315206684 (ebook)
Subjects: LCSH: Systems engineering—Simulation methods. | System design.
Classification: LCC TA168 .M66 2019 | DDC 620/.0042—dc23
LC record available at https://lccn.loc.gov/2018027840

Visit the Taylor & Francis Website at
http://www.taylorandfrancis.com

and the CRC Press Website at
http://www.crcpress.com

Contents

Part 2 Network-Centric Operations and Matters of Context

Part 3 Organizing in Business Networks

Part 4 System Operation and Configuration Management

Preface

Modern engineering systems such as an aircraft or a frigate are complex and multifaceted. They must be flexible and adaptable to suit the environment of individual mission requirements, and fully integrated with the supply chain and other stakeholders to deliver an ongoing satisfactory level of performance. In essence, these systems are working within a larger socio-technical system that is governed by a different set of rules and operates with a different type of structure. Engineers trying to apply the theory of systems engineering to "design" a system of systems can find the dynamic nature of connectivity and unexpected interactions perplexing. We draw parallels with natural ecosystems, where changing the balance can have unexpected effects.

The study of system of systems incorporates concepts and practices in many dimensions. Firstly, the traditional systems engineering lifecycle management theory is an essential constituent in the design, development, and control of human created systems. Secondly, the processes and procedures are managed as soft systems that can be changed, modified, and refined. Thirdly, operation of the system of systems depends on the understanding of people and roles in the system. Last but not least, these systems are constrained by the environment in which they are operating. The environment includes regulatory and institutional requirements. Many of these constraints are not visible until end of the system's lifecycle.

This book addresses the skill set of professional engineers and operations managers required to design, develop, implement, and operate a complex socio-technical system encapsulating many engineering systems. This book provides a core knowledge base that draws upon principles derived from a wide range of business and engineering disciplines including systems engineering, services engineering, operations support, integrated logistics, industrial engineering, and the outcomes of the latest network and holistic system research.

The book is intended to be both a special text for relevant courses at university level as well as a general knowledge textbook for professional engineers, systems managers, and operations managers to enhance their knowledge of complex engineering systems and their integration into a system of systems. The book is also written with the aim that research students studying in systems engineering, support and services engineering, logistics, and procurement will find it useful as an introductory text to start their research aspirations.

This book has two major knowledge themes on system of systems. The engineering of system of systems concerns the knowledge required to design, plan, and analyze a system of systems prior to and during its development. Operations of system of systems concerns the knowledge required to manage, use, and implement systems of systems while serving the needs of the stakeholders. These engineering and business viewpoints of expensive, complex engineering systems are addressed in several ways:

- Creating an operational view and new understanding of modern system design, commissioning, operation, services, and support. Contemporary books on systems engineering only focus on the physical boundary of the system being analyzed, and this restricts the ability to manage across socio-technical system boundaries.
- Including system of systems modeling and analysis techniques that are essential for engineering professionals and business development managers to develop a whole-of-system view of the requirements of a system of systems for its serviceable business life.
- Taking a system of systems approach incorporating systems engineering principles and network-centric concepts to guide the reader through the important development processes of design, development, testing, and verification for supporting complex system of systems engineering and operations in a changing business environment. This approach is different from many engineering management books that only focus on management of the engineering process rather than extending the process to other systems in the constraining environment.
- Providing a detailed account of performance assessment and risk analysis methods that are critical to support changes of the systems and their network. Systems are changed because the internal or external variations during the operation lifecycle of the system. The system of systems is therefore changed as well due to external environmental influences.

The book uses case studies and worked examples to illustrate concepts and principles. It is organized in four parts, each with a collection of chapters on particular aspects of system of systems.

Part 1, Systems Thinking, focuses on the development of fundamental concepts of a system, how it is viewed from within the system, and how a system operates in an environment where many other systems exist.

Chapter 1. Foundation of Systems Thinking

This chapter introduces the fundamental principles of systems engineering and the methods and tools for supporting systems engineering analysis. The practice of system thinking is reviewed and refined. In addition, the importance and effects of system integration and whole-of-life systems design are highlighted with examples and industry cases. Using a medical example, this chapter illustrates the foundation of systems of systems from systems engineering concepts and principles.

Chapter 2. Defining System Architecture

This chapter explains the needs and practices of applying system architecture to understand existing complex systems and to design and implement new complex systems. Standard architectures such as GERAM, GRAI, Zachman, DoDAF are explained. The system architectures are explained in the best context of application that these architectures are developed for.

Chapter 3. System of Systems Framework and Environment

This chapter focuses on the International Standard ISO 42010. This standard defines terms, principles, and guidelines for the consistent application of architectural precepts to systems throughout their lifecycle. The key functional blocks—architectural description, stakeholders, and rationale—are examined in relation to their effects on interacting networks. In this chapter we bring functional system and model views together with a focus on designing dynamically reconfigurable systems of systems.

Chapter 4. Modeling of Socio-technical Systems

This chapter covers the theory of a generic enterprise model that contains the conceptual elements of product, process, and people, and their interactions within an operating environment. The modeling method is then extended with quantitative analysis, and demonstrated by analyzing changes in a company going through an internal improvement process.

Part 2, Network-Centric Operations and Matters of Context, expands the perception of infrastructure to a much broader concept. Within an infrastructure, many systems are interacting among themselves and there are unspoken conventions that are adapted through networking activities. To understand the complexity of this system of systems, the network modeling approach is introduced. We also discuss the impact of the global communication mechanism—the internet.

Chapter 5. Infrastructure Network Applications

In this chapter, the concept of infrastructure is extended from the usual meaning of buildings and roads to include organizational structures with financial, information and knowledge components, and consider their combination with physical components. Infrastructure of any kind may be viewed as a system of systems, with vulnerabilities associated with the weakest link or coincident failures of some sort in more than one subsystem.

Chapter 6. Network-Centric Operations

Network-centric is a term often used in a military context, but it is now increasingly used in commercial operations. A network-centric organization has a particular way of thinking and working in a complex operating environment. In this chapter, some common themes that affect how they are designed and operated as systems of systems are explored, particularly in the context of organizing to manage emergent events.

Chapter 7. Social Networks and Creative Collaboration

The concept of actors and activities is introduced to explore connectivity between different elements in a system of systems. Social networks are an element of social capital that an enterprise or community may draw upon. Social networks allow systems in the same environment to connect and share. The design and operation of complex systems of systems requires the cooperation of people or organizations, and access to resources that may not be owned by a particular enterprise. This chapter highlights the factors to be considered when establishing collaborative working arrangements.

Chapter 8. IT Infrastructure and the Internet of Things

Over the last two decades, the proliferation of the Internet of Things has changed the way global business networks are operated, planned, and controlled. The internet offers businesses enormous potential and entirely new methods for streamlining coordination between business partners, improving service quality and flexibility, and reducing operating costs. This chapter reviews how the internet is used in business and engineering environments, and examines the major enabling technologies of the internet-enabled global business network.

Chapter 9. Governance

This chapter outlines the many constraints that are inherently imposed on a system irrespective of its nature and how it is formed. A system with good governance provides an environment for the effective and efficient creation, promulgation, and implementation of essential regulations to meet the aims of regulating authorities.

Part 3, Organizing in Business Networks, outlines four essential concepts for system of systems. The study of virtual enterprises started in the 1990s, before the internet existed in a form that could enable close collaboration. As information and communication technologies have advanced since the millennium, the importance of service operations in business networks has been recognized and researched. As opposed to product developments, service developments depend on innovative ideas and their implementation, and this concept is examined in the last two chapters of Part 3.

Chapter 10. Virtual Enterprise Collaboration Concepts

A virtual enterprise is an evolutionary business entity that represents several states of a system of systems, from a virtual breeding environment to project collaboration. In the global business environment, product development and knowledge management is critical to the success of a company. This chapter explores the nature and characteristics of virtual enterprises. This provides an understanding of how business is changing by the application of the latest information and communication technologies.

Chapter 11. Markets and Service Ecosystems

In this chapter, drawing on systems architecture framework concepts and considering the myriad of systems of systems associated with a market perspective, the idea of service-dominant logic—promoted by some marketing and service providers—is introduced. As markets have strong focus on delivering value to customers, they operate on information and business intelligence that transform organizations to comprehensively embrace digital technology.

Chapter 12. Business Models Making Operational Sense

The design and development of any system of systems has to make both operational and business sense. This can be achieved by linking operational and business model perspectives. This chapter introduces business modeling and how design-thinking principles can be applied to facilitate business model innovation.

Chapter 13. Innovation in Ecosystems

Innovation is revolutionary. The ecosystem drives significant change in a system in many ways. The environment, which includes a range of social and economic factors, is changing all the time. These changes are external to systems and often are not anticipated in gradual development plans involving continuous improvement cycles. It is important to develop innovative thinking to build the ecosystem and firm up the main achievements during the innovation process.

Part 4, System Operation and Configuration Management, outlines methods that can be used to describe the characteristics of a system of systems, and to measure its effectiveness. The models developed from these tools can then form the baseline for the analysis of more complex system constructs.

Chapter 14. Capability and Performance Assessment

Servitization is a term used to describe the way that companies making complex engineering products are extending their businesses to offer services and support after their product is commissioned. Performance-based contracting is a recent business model in which revenue depends on the achievement of satisfactory performance indicators. Significant risks to service providers exist. This chapter explains how the 3PE model presented earlier can be used to assess the capability of an organization to fulfill a performance-based contract.

Chapter 15. Managing System Models

Generic system modeling methods and tools are useful for a standard system environment. This chapter describes several key system modeling tools that supplement the 3PE system model to represent different aspects of a system of systems. Once a baseline system model is established, it can be used within the 3PE modeling framework for analyzing interrelations with other systems and overall system of systems characteristics.

John P. T. Mo

Ronald C. Beckett

Authors

Professor John Mo obtained his PhD in Manufacturing Engineering from Loughborough University. He is professor of Manufacturing Engineering and a former head of the Manufacturing Engineering Discipline of the School of Engineering at RMIT University. He has expertise in systems integration, data communications, and sensing and signal diagnostics. Prior to joining RMIT, he was a senior principal research scientist at CSIRO and led the manufacturing systems and infrastructure network systems teams, among others, in the Division of Manufacturing and Infrastructure Technology. Over 11 years in CSIRO, he led a team of 15 professional research staff worked on risk analysis algorithms, electricity market simulations, wireless communications, fault detection, and production scheduling. He was the project leader for many large-scale government projects, including productivity improvement in the furnishing industry and consumer goods supply chain integration.

Professor Ron Beckett is an Engineers Australia Fellow, and holds a Doctor of Business Administration (UWS) based on research into ways of implementing organizational learning concepts in a high-technology company. He has more than 25 years of experience in R&D, operations, and strategic management in international aerospace manufacturing, and more than 10 years of experience in management consulting. He works at the academia-industry interface and has written more than 100 papers related to innovation, collaboration, and knowledge management. He is an adjunct professor in the Swinburne University of Technology Business School, and lectures in some Masters of Engineering subjects at RMIT.

Part 1

Systems Thinking

1

Foundation of Systems Thinking

> A system inhabits an environment. A system's environment can influence that system. The environment, or context, determines the setting and circumstances of developmental, operational, political, and other influences upon that system. The environment can include other systems that interact with the system of interest, either directly via interfaces or indirectly in other ways. The environment determines the boundaries that define the scope of the system of interest relative to other systems.
>
> **IEEE Architecture Working Group, September 2000**

1.1 The Role of Engineering in Systems

According to the International Council on Systems Engineering (INCOSE), a not-for-profit membership organization, "systems engineering" is an interdisciplinary approach and methodology for the realization of successful systems. It focuses on defining customer needs and required functionality early in the development cycle, documenting requirements, and then proceeding with design synthesis and system validation while considering the complete set of problems in the application area. In short, systems engineering integrates all engineering disciplines and specialty groups into a team effort by a structured development process to achieve the desired systems development goals.

Through the systems engineering process, the systems engineer controls the use of resources in the form of materials, machines, equipment, and competency (the people) to create an engineering system that fulfills human wishes and needs. Systems engineering draws on all relevant knowledge to create the system and applies the information characterizing the system so as to inform the end users what the system is and how it works.

Nature is governed by a universe of principles that have been or are being discovered by scientists. Systems that are developed for the benefit of human race are created by applying relevant scientific principles to produce desirable outcomes. We'll examine a few systems, from simple to complex. We'll identify and illustrate how the scientific principles involved are applied to these cases.

1.1.1 Engineering of Traffic Lights System—Governed by the Science of Braking

The traditional traffic light system works in conjunction with other systems, viz, the road system and the vehicles (which themselves are combined systems of the drivers and their vehicles).

In Figure 1.1, the yellow light (indicated by light gray) is a warning to stop. In practice, the driver should move to brake mode so that the car can stop before the light turns to red (dark gray). The system here is not limited to the traffic lights—it also includes drivers and cars.

How much time should the system allow for the yellow (light gray) light to stay on?

Using the variables on the diagram, the time t for the vehicle to stop before the stop line (double bold lines) is governed by the following equation:

$$\frac{ft^2}{2} - v_0 t + x - x_0 = 0 \tag{1.1}$$

where x is the instantaneous distance and f is deceleration caused by applying the brake. Using some reasonable numerical values for Eqn. 1.1,

- For a car with m = 1,000 kg, v_0 = 60 km/h, braking force = 10,000 N, time to stop is 4.3 s
- For a truck with m = 30,000 kg, v_0 = 60 km/h, braking force = 100,000 N, time to stop is 11.1 s

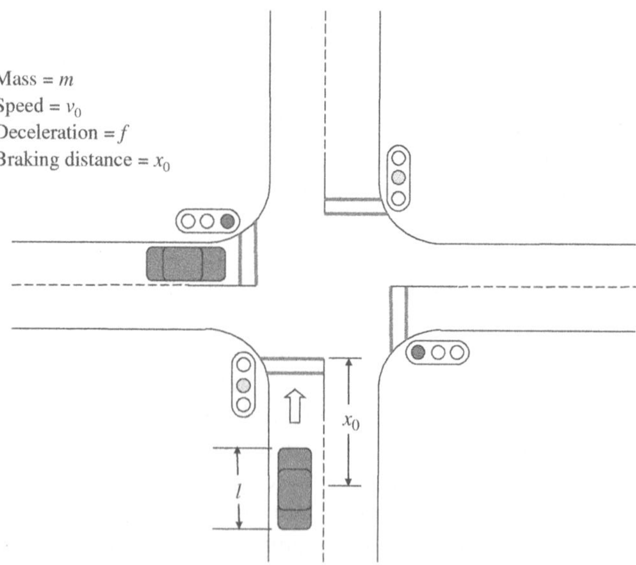

FIGURE 1.1
Traffic light system

Hence, in deciding the "on" time of the yellow light, we need to consider all elements in the system, in this case the stopping time of at least 11.1 s.

This time setting within the traffic light system not only affects the cars and trucks around the junction, it also affects congestion, travel time, and at some locations, air pollution around the junction.

1.1.2 Engineering a Safe Road System—Based on the Science of Statistics

In a rural community, many road sections are marked with a central double line, indicating that no overtaking is allowed. However, many drivers just ignore this regulation and overtake, even on rainy days when visibility is poor and stopping distances are increased. In the scenario shown in Figure 1.2, how can the road system prevent a tragedy from occurring?

In this case, the science of statistics can be applied to develop the system. The attitude of drivers depends on a lot of factors. A road system is not designed to change the psychology of drivers. Hence, in a road system, even if only one driver takes the chance, there is a probability for a tragedy to happen. Decision on investment to design a safe road systems depends on statistics. Statistics show that there are more deaths on the road in rural areas than in urbanized areas. For example, Table 1.1 shows that nearly two-thirds (65%) of fatal accidents in Victoria, Australia, occurred in rural or provincial areas, where the road systems are not engineered to high safety standards.

FIGURE 1.2
Road system

TABLE 1.1

Road Toll Statistics in Victoria, Australia

	2014 Road Toll	2015 Road Toll
Provincial cities/towns	17	18
Rural roads	139	142
Small towns/hamlets	4	3
Urban Melbourne	88	89

A safe road system should ideally have no fatalities. The statistics in Table 1.1 indicate that rural roads (which are mostly bi-directional) are the most dangerous. The systems engineering principle for a safe road system is physical separation of the traffic traveling in each direction. If there is strong guidance to road users, this type of irrational action cannot take place. A secure central barrier, as shown in Figure 1.3, can enable traffic to flow smoothly and safely.

FIGURE 1.3
Safe road system

A well-established safe road system consists of a number of components:

- Efficient highways
- Freeway interchanges
- Strict licensing regulations and management
- Public education on road safety

This system not only enables good traffic management, but also benefits society by reducing accidents, improving travel times, and lowering compensation and remedial costs.

1.1.3 Underground Train System—A Large Engineering System Based on a Range of Scientific Principles

How can an underground train system be created?

In London, the underground train system dates back to 1863 (Figure 1.4). Since then, the system has been continuously developed and is still being extended today. The network has become very complicated and yet it is

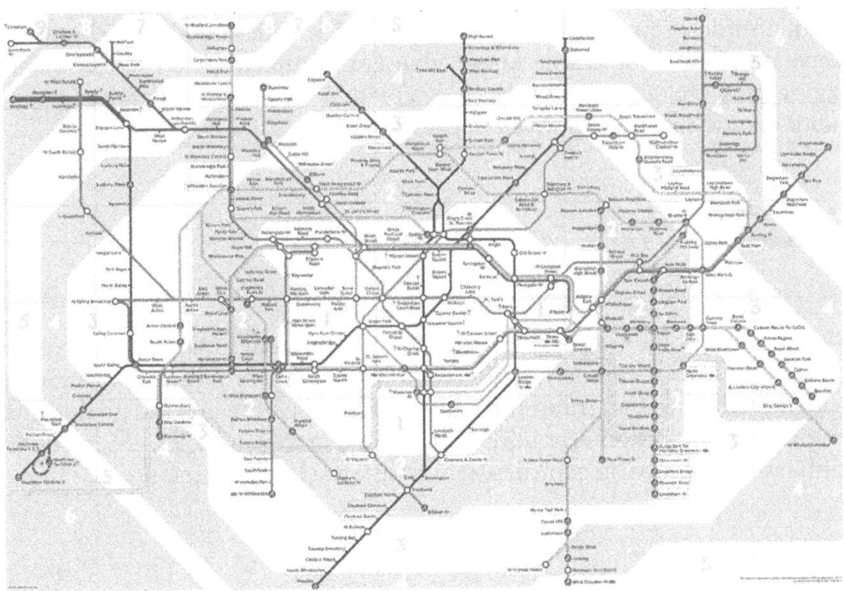

FIGURE 1.4
London underground transport system

(Image captured from document: http://content.tfl.gov.uk/standard-tube-map.pdf, 2016)

regarded as one of the safest transport modes in London. The system consists of many parts:

- The "Tube" tunnels and air flow system
- Rolling stock (trains) and tracks
- Power system
- Signaling system
- Ticketing system
- Management organization

Over the last 150 years, as the London underground train system has been developed, many scientific discoveries have been made. Engineers have applied this new knowledge to expand the system and improve those parts of the system that have already been built. For example, tunnels are now bored at greater depths with better understanding of the underlying geology. The signaling system has been revamped from the hand signals used initially to the digital systems we know today. The rolling stock (i.e., the trains) are designed to be much safer and more comfortable. The scientific principles applicable in this case are materials science, heating, ventilation, and cooling principles, reliability assessments, etc.

When developing a sophisticated engineering system, it can take time to transform current scientific knowledge into practice. In many cases, like the London underground system, we have to work with less ideal technologies when the body of scientific knowledge is incomplete. The role of systems engineering therefore is not only to create a working system but also to maintain and improve the system over time.

1.2 System Challenges

A system is a unified set of components with different functionality working together toward a common goal. Not surprisingly, bringing many components together and aligning the varying functions toward the goal is challenging. Systems engineers have to overcome many obstacles, e.g., resolving conflicts or reinforcing coordination, when designing systems.

1.2.1 Integrity

Engineers have a responsibility to create, design, manufacture, manage, and dispose of systems that operate safely, reliably, and with minimal negative impact to the society. Human lives can depend upon the quality of engineering projects, and significant economic and environmental consequences can result from underperforming facets of the system. Therefore, when designing systems, engineers should always be aware of the dangers and limitations of systems irrespective of what benefits the system can bring.

The challenge of integrity in systems is to maintain total knowledge of the principles, characteristics, constraints, and processes that exist around the engineering system so that any foreseeable problems can be prevented and any damages can be minimized, even in extreme circumstances. A good example is the International Space Station, which has been created with system integrity as the first principle to ensure it is a liveable space orbiting at an appropriate altitude around the earth, i.e., all parts of the system must work toward this goal all the time.

1.2.2 Stability

The challenge of stability originates from mechanics and structures. When a mechanical structure is stable, all forces in the structure are in equilibrium so that loads are distributed to structural members that can bear the load for a suitably long time. Similarly, in electrical systems, a stable electrical circuit is one that has an equilibrium of voltages and currents being distributed appropriately.

In all systems, the maintenance of equilibrium is necessary to ensure that the system can operate and perform at the right level for a long period. The International Space Station was created to support scientific experiments and other activities over time. Stability of the system and confidence that it can be maintained in this state for a predictable period is critical to its success.

1.2.3 Compatibility

A compatible system is one that can exist and operate in a harmonious, agreeable, or congenial manner with other systems. Compatibility can be achieved in many ways. As an example in a biological or medical system, a cochlear implant should be able to exist in the body in a chemically and biologically stable state, and work harmoniously with other parts of the body—in other words, to be compatible. In addition to existence, the cochlear implant must also perform in a manner that transmits understandable information to the brain.

Likewise, in an engineering system like the International Space Station, the need for compatibility arises from the existence of shared resources and usages. It has to be compatible with the approach of spacecraft of different makes. It has to make allowance for operation by different teams and different sets of equipment. Simple things like the inability to fit a screw into a different threaded helix in an item made in another country can ruin a mission. The challenge to ensure compatibility should not be underestimated.

1.2.4 Safety

It is well known that a construction site is full of danger. Many work accidents in such locations are reported every year. This is why stringent regulations are in place to ensure the safety of workers on site, such as for the wearing of helmets and the posting of clear signage. In this case, the challenge of safety is manageable because the environment is generally known, predictable, and controllable.

Designing and building a new system in an extreme environment can be more challenging. Any minor error can easily escalate to disaster if not carefully managed. When designing such a system, many safety measures must be included, and processes must be defined and rehearsed to ensure these safety measures are followed. Each of these measures can add another layer of protection on a number of potential dangerous events if any such event occurs.

1.2.5 Sustainability

A large, complex engineering system requires huge investment from stakeholders. It is clear that such a system is not intended to be used just for a short time. These kinds of systems are typically expected to be in service for 30 years or more. Thirty years approximates to the length of a generation, so it is common that such complex systems remain in operation for several generations. During this time, many changes can take place. For example, technology may change so that the existing system becomes incompatible with more modern coordinating systems, or some components wear out after many cycles of operation (the so-called ageing effect).

These changes can also happen within the expected service life. Sustainability must be designed into the system, so that appropriate maintenance and upgrade services can be carried out in a timely fashion. Part of a system's sustainability design entails being prepared for the performance of these services. A good example of this challenge is the Hubble Space Telescope. After its initial commissioning and launch, it has had to go through five more repair and upgrade missions, making the overall cost of the system very high. The challenge of sustainability is to design for these unexpected but preventable changes to occur so that in case there are any incidents that affect the ability of the system, a more controlled and better managed process can be put in place quickly and economically to resolve the problems.

1.3 Case Study

The cochlear implant is a device based on the scientific discovery of a number of acoustic and medical principles.

The brain receives external auditory information via electronic signals generated by sensing cells in the ear. For a variety of reasons, this natural system fails in the hearing impaired. A cochlear implant generates electronic signals when sound waves reach the external sensor. The implant transmits these signals directly to the nerve endings in the cochlea by means of an electrode array and associated components surgically inserted into the inner ear or just below the skin.

The external sensor is a speech processor that sends electrical signals to the implant in the form of radio waves. It looks like a hearing aid but the structure and functions are significantly different, and more sophisticated.

Many people from different fields and occupations have contributed to the design, development, and implementation of this system. There are many components involved, including the speech processor, the implant, and the radio wave transmitter.

However, a cochlear implant is not a natural part of the human body. The electrical signals it produces are different from those produced by the inner ear. The patient needs training to adapt to the sound wave pattern it generates. The system also requires occasional tuning and maintenance, and it is not unusual for the patient to receive additional clinical treatment months after the implantation operation. Hence, in addition to the hardware components described, the "system" in this case also includes the people who must apply their own specialized knowledge to enable different parts of the system to work together.

1.4 Systems Engineering Lifecycle Model

At this point, it is important to clarify that the systems referred to in this book are artificial systems, i.e., designed and built by people for a variety of purposes. Natural systems occurring and developing spontaneously are out of the scope of this book.

While some systems may be developed in a haphazard or unplanned way, most systems are designed and built by the systems engineering development process. This is essentially the left-hand side of the V-model systems engineering lifecycle model shown in Figure 1.5.

This chapter reviews the fundamental principles and key features of this part of the lifecycle. We refer to this work as the "engineering" of the system. The term "engineering" in this context involves three actions:

- Analyze—Systems should be designed to be fit for purpose. To ensure that this goal is achieved, systems engineers must analyze many things: what the user actually wants, what is affordable, what stresses the system will need to overcome, what sequence of operations are expected in reality, and so on.

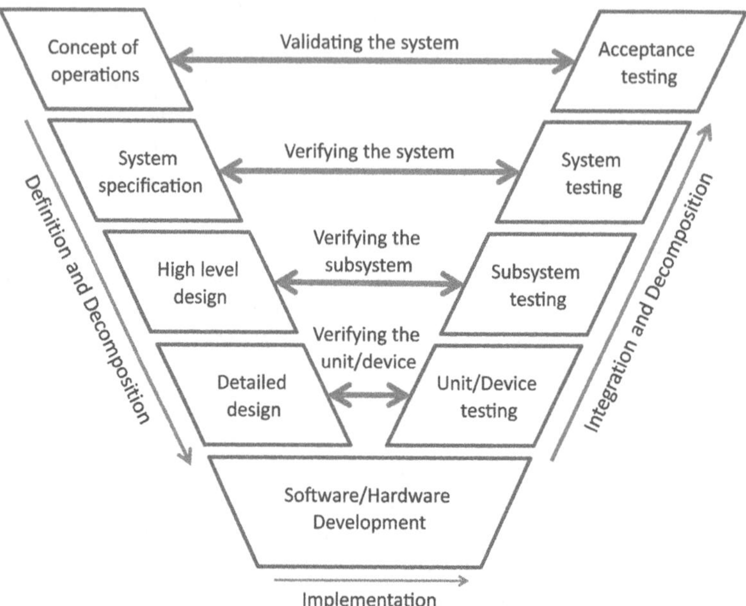

FIGURE 1.5
Systems engineering V-model lifecycle model

- Design—Based on the user requirements, systems engineers must balance the design with other constraints imposed on the system. For example, cost is usually a key constraint to be considered. Other constraints can include the time available, the labor required, etc.
- Plan—Design activity specifies the ad hoc look and feel. However, turning the design into the real system requires planning of the steps executing. Activities that are relevant to the preparation of the project could be specified and the resources can be scheduled according to time.

On the right-hand side of the V-model is the verification of the system. This involves four actions:

- Device testing—Realization of the system requires the actual building of the system as a combination of hardware and software. However, except in the case of very small systems with just a handful of components, we normally build systems with many devices. These devices can be developed individually and they need to produce effects (i.e., perform) as they are intended. These components are often purchased from suppliers who specialize in producing such devices in a mass-production facility. The function of device testing is to prevent the acceptance of faulty devices, which could render the whole system useless.

- Subsystem testing—A large system is often composed of subsystems, that can be activated to produce certain effects. In a sense, these subassemblies are more complex devices created and manufactured by the engineering project team. Subsystem testing is intended to make sure they will perform as they are designed.
- System testing—After all devices and subsystems have been successfully tested, the system can be assembled. Integration of a wide variety of components of different forms and behaviors involves uncertainty and can be problematic. The system testing process can take months and requires many different operating scenarios, depending on the concept of operations as specified by the user.
- Acceptance testing—If everything goes well, validation and acceptance of the system by the customer should be a routine step. This means it is better to delay the validation run until the system performs perfectly in system testing.

1.5 Reflections

From your experience working with an engineering system either at work (e.g., on a bridge, at an airport, in a factory, on the internet) or at home (e.g., your smartphone, your gas stove, your washing machine), answer these questions:

- What are the scientific topics that apply and contribute to your system?
- For each knowledge topic in the answer to the first question, list the "engineering creation" required to transform that knowledge into usable components that form part of the system.
- What is the role of engineering in the creation your system?

Discuss which branches of engineering (in addition to systems engineering) are required to transform the system from design to the "real world".

1.6 Additional Reading

Elliott, C. & Deasley, P. (Eds.) *Creating systems that work: principles of engineering systems for the 21st century*. Royal Academy of Engineering, 2007. ISBN: 1-903496-34-9

2

Defining System Architecture

> With the globalisation of economies, enterprises are operating as large, complex networks of autonomous units. Design offices, manufacturing facilities, services and maintenance stations are scattered around the world. Component supplier networks provide vital subassemblies and parts. Such intricate field of endeavour needs a clear operational structure in order to design and manage the continuous metamorphosis necessary not just to keep businesses alive, but also to strive for growing businesses and operate them successfully during their whole of life.
>
> **Peter Bernus and Laszlo Nemes**
> *Handbook on Enterprise Architecture*

2.1 Why Do We Need Architecture Modeling?

A model is a representation of the real world. There is no right or wrong model. However, a model may be distorted by the process of developing it, and may no longer resemble reality. In this case, the model is said to be a bad model. On the other hand, if the model produces a meaning which is close to the real world, the model is said to be a good model. An architecture provides a basis for generating the model. Architecture modeling helps the designer to focus on key features so that the chance of developing a useful model is greater.

2.1.1 The Objectives of Architecture Modeling

The word "architecture" commonly refers to the structure and arrangement of buildings. In a more generic sense, architecture has to do with planning, designing, and managing space to exploit functional, technical, social, environmental, and aesthetic applications. It requires the creative manipulation and coordination of materials and technology, and of light and shadow. Figure 2.1 shows a functional space with a consideration of aesthetic elegance, making it unique within its social and urban environment. The creation of this architecture drew upon previous knowledge of building on a relatively small footprint and on slopes. The knowledge that has been captured in this architecture includes designing for the provision of utilities, emergency access, and earthquake resistance.

FIGURE 2.1
An illustration of a building architecture

An architecture is not a blueprint. It forms the basis for design and planning. In Figure 2.1, you may notice there are two buildings behind the main building with a similar architecture, probably built around the same time. Their basic features are largely identical but with minor modifications as might be expected in a different location.

In this context, an architecture has some typical characteristics:

- It captures previous experience in a knowledge structure.
- It provides a baseline to build a new structure.
- It represents the minimum expectations.
- It offers guidance for the work of the team.

In systems development, an architecture is not the same as a development standard, but it serves a similar purpose, providing structure around design decisions and processes. System architecture deals with the structure and arrangement of systems. A system is not just the implementation of hardware and software—it has meaning in the society, environment, and business in which is it located and employed.

Using a more commercially oriented term, a system can include sociological forms such as companies, government and non-government organizations, project teams, collaborative units, or functional groups, for example. From an operating point of view, a system is an enterprise that has a mission, and it is in the interest of everyone involved in the enterprise to support the activities, and the utilization of resources, being put towards the mission.

An enterprise is not a naturally evolved phenomenon. It needs to be formed within society. To assist the formation of enterprises, enterprise modeling is used to take some standard templates of enterprises and use them to plan for its development. The objectives of enterprise modeling are:

- Integration of the enterprise business process
- Integration of the enterprise information system
- Facilitation of effective communication and interaction among the enterprise participants
- Provision of support for sound decision-making

There are a lot of benefits derived from a well-planned enterprise. Tangible outcomes include improved product quality, decreased unit cost, and reduction of development cycle time. Non-tangible benefits include increased flexibility, ease of access to information, and the ability to monitor and provide automated, encouraged cooperative work, and globalization.

2.1.2 Modeling for Transition Planning

The main reason for developing a system model is to plan for the future. When creating a system, the engineering design team must first establish a baseline against which the future can be benchmarked. To represent the baseline, depicted as the current scenario (AS IS) in Figure 2.2, system modeling tools are used to describe the present. User requirements are captured and explored to create the future scenario (TO BE). The future scenario

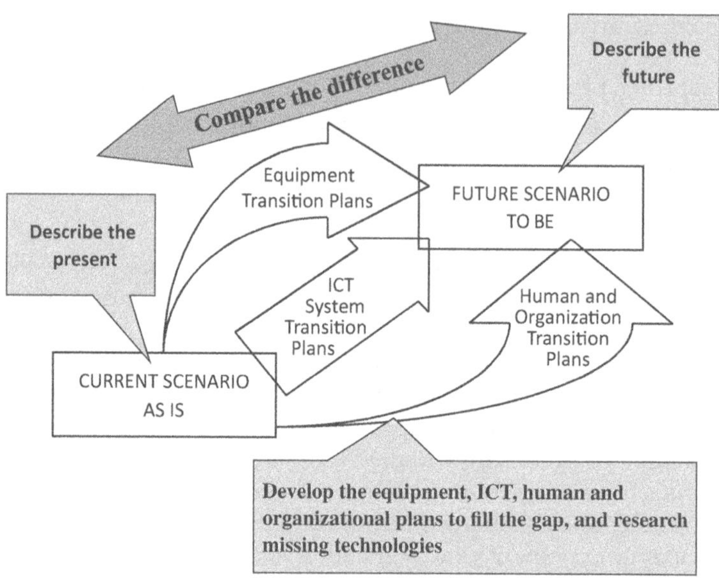

FIGURE 2.2
Transition planning

should be described with the same system modeling tools used for the baseline. In this way, the engineering team can set its objectives. Figure 2.2 shows the relationships of system models in transition planning.

Using the same set of system modeling tools, comparison of the features of the AS IS and TO BE scenarios is possible. With the new elements of a system defined, plans to migrate the product (equipment), people (human and organization), and processes (ICT rules and procedures) to the future state can be developed.

In the following sections, we'll examine some well-known enterprise architectures that have been used to guide the development of enterprise systems and large engineering projects.

2.2 Enterprise Reference Architecture

2.2.1 CIMOSA

The European Strategic Programme on Research and Development for Information Technology (ESPRIT) was a large-scale research program funded by the European Union in the 1980s. The first ESPRIT program, resulting in 226 projects, cost 1.5 billion euros while the second ESPRIT program cost 3.2 billion euros. The Computer Integrated Manufacturing Open Systems Architecture (CIMOSA) was developed under ESPRIT Project 688 by the

AMICE Consortium with several other participants, including Aerospatiale, GRAI, Hewlett-Packard, ENSIDESA, NLR, AT&T, FIAT, Italsiel, Alcatel, Capgemini, Digital, Daimler-Benz, IBM, ICL, Siemens, WZL RWTH Aachen, and others.

The CIMOSA project studied a wide range of problems in the manufacturing industry in Europe:

- Trade barriers/open market competition
- Worldwide availability
- International competition
- Change management
- Collaborative investments
- Flexibility in systems
- Efficient use of resources
- Information compatibility
- Software development and maintenance
- Reusability
- Integration of dissimilar systems
- Multiple vendor systems

The objectives were to create an architecture with these features:

- Enables enterprises to operate in real-time adaptive mode
- Provides architectural constructs for computer integrated manufacturing (CIM)
- Supports multi-disciplinary knowledge integration
- Provides a complete description for IT applications

The outcome of the CIMOSA project can be represented by the building blocks in Figure 2.3. In fact, many standards announced in the 1990s and around the millennium have their origins in these early "scientific" enterprise architectures.

CIMOSA relies on three evolutionary processes to develop the final system architecture for an application. These three processes (generation of views, instantiation of building blocks, and derivation of models) are represented by the thick arrows outside the blocks.

The first process is used to generate four views: function, information, resource, and organization. These views represent the essential elements of any enterprise. "Enterprise" here can mean a company, a system within a company, a project, an automated system, etc. Effectively, the function and information views reflect the system process, the resource view reflects the product and the organization view reflects the people.

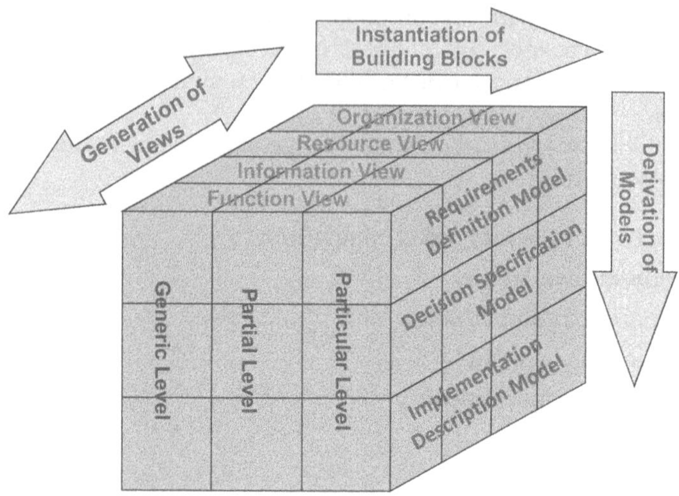

FIGURE 2.3
CIMOSA basic building blocks

These views represent what is required within a system. The vertical axis represents the systems engineering design process that we discussed in chapter 1. The three models are milestone phases in the development of a final system that meets user requirements, i.e., requirements, design, and implementation.

What makes CIMOSA unique is the instantiation of building blocks. There are three levels: generic, partial, and particular. The **generic level** contains blocks of views (horizontal) and models (vertical) that can describe systems of almost any industry. These are the fundamental characteristics common to any system that would be used when building a system from the ground up. However, the generic level leaves a lot of room for different interpretations if not further elaborated. The **partial level** building blocks are elaborated for a particular industry, e.g., automotive, IT, aerospace, utility. Each industry has specific requirements for function, information, resource, and organization. Expanding on the industry-relevant requirements gives a better set of building blocks for implementation within that industry. The **particular level** expands further on the specific needs of a particular company. The building blocks defined at partial level are already sufficient for most companies but some special requirements may still exist for an individual company to consider. Once the final building blocks are defined, the system can be built.

2.2.2 PERA

The Purdue Enterprise Reference Architecture (PERA) was developed from earlier works in the Purdue Laboratory for Applied Industrial Control at Purdue University, such as the steel and paper industry Hierarchical Control

System studies, the Purdue Reference Model for CIM and the Industry-Purdue University Consortium for CIM.

Having no method for writing algorithms defining the human-generated innovation involved in the functions in processing systems, the CIM Reference Model committee chose to consider human-based functions as external entities (i.e., outside the sphere of the model). As long as one focuses solely on control system tasks (most of which can be automated), it is only necessary to supply needed plant information, and receive instructions and data from the human side, and the model is then as complete as necessary.

These problems were solved by defining a generalized task representation which covered information system tasks (the algorithmic control tasks of the CIM Reference Model) and manufacturing tasks, as well as the human-based tasks.

Figure 2.4 illustrates PERA. This system is characterized by the layering of the lifecycle diagram into task phases covering the full lifecycle of the enterprise involved.

Although PERA was developed for the steel and paper industries, Figure 2.4 shows that it is quite generic, in that there is little reference to the specific industries in which it was researched. Effectively, PERA is a time-based enterprise architecture representing the evolution of activities, resources, structures, and information in the life of the enterprise. The layering of tasks occurring within those regions is represented graphically as a top-down approach. Hence, the resulting architecture could apply to any enterprise regardless of the industry involved, and its applicability was far beyond what was originally intended.

2.2.3 GRAI Integrated Methodology

The work on Graphs with Results and Actions Inter-related (GRAI) methodology started in the 1970s at the GRAI Laboratory of the University of Bordeaux. The objectives at that time were to model a production management system in order to be able to define precisely the specifications needed to choose a software package for a computer-aided production management system.

The first application of the GRAI method for production management was made in 1980 for French energy company Télémecanique électrique. Between 1980 and 1985, 30 applications of the GRAI method were realized either by the GRAI Laboratory itself or by other engineering firms. In 1985, the GRAI method was used by aircraft engine manufacturer Snecma to design and specify the production management system of its automated factory in Le Creusot, France. In the late 1980s, the GRAI method was expanded into an enterprise integration methodology, called GRAI Integrated Methodology (GIM).

The GRAI models provide a generic description of a manufacturing system while focusing on the details of the control parts of this system. GIM extends the GRAI concept to the control of a manufacturing system being represented from a global point of view through levels of decision centers. Figure 2.5 shows the key features of GIM.

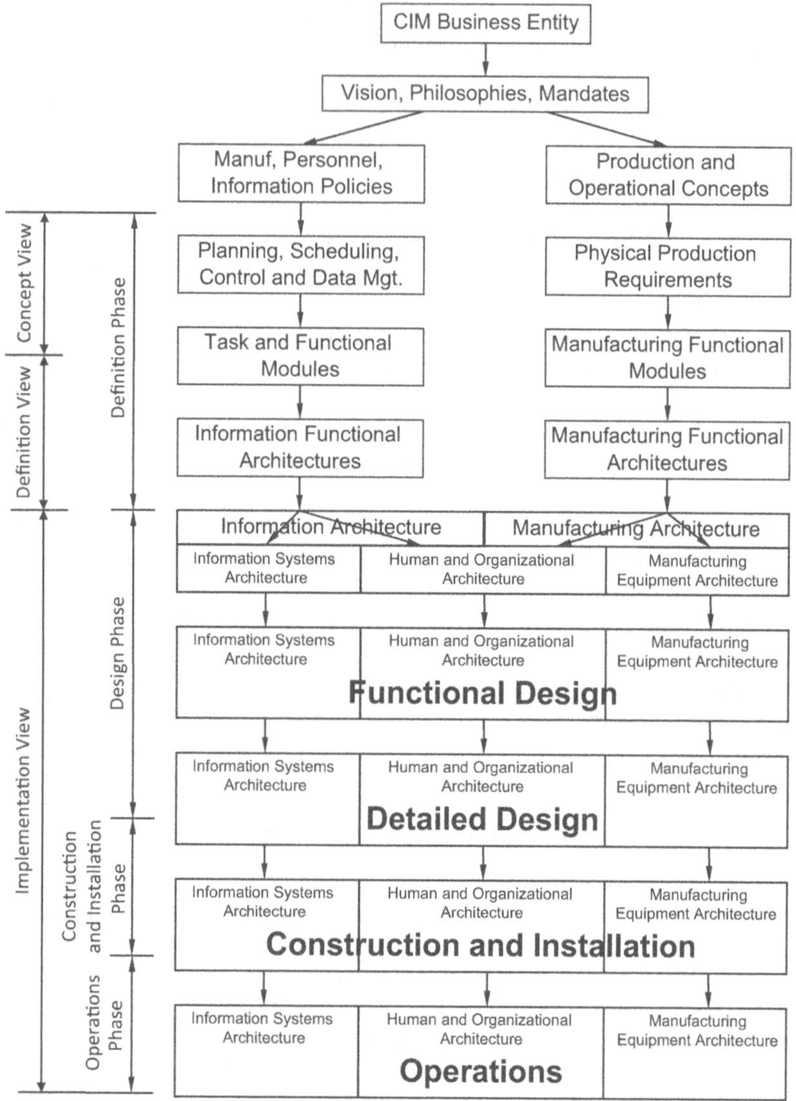

FIGURE 2.4
PERA phases

From GIM's perspective, the physical system represents all the hardware investments that have been deployed to produce the required outcomes. The physical system is controlled by a number of decision centers. Most of these are at the level immediately above, but some decisions may be made from decision centers at higher levels. The decisions indicate a defined hierarchy of control, which is normally recognized in the organizational structure.

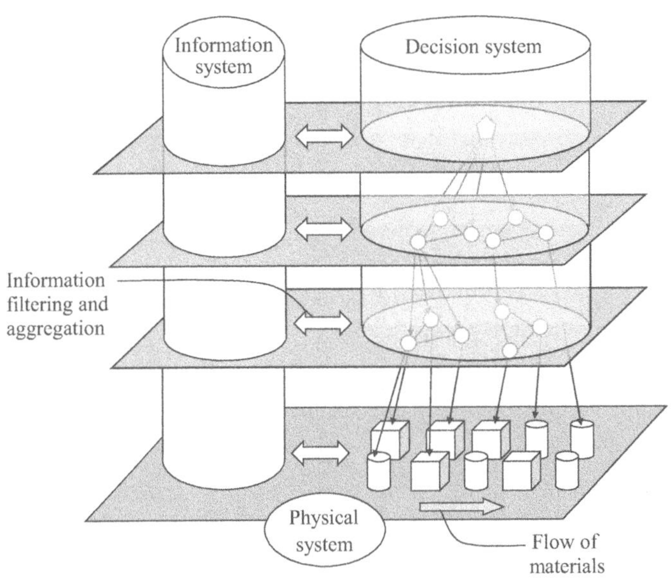

FIGURE 2.5
GIM control by decision centers

However, there are occasions when some decisions might be made outside of the formal structure. The GIM model captures both types of decisions.

To assist the decision centers, both for the information required to make decisions and for communication between levels of decision centers and the physical system, an information system infrastructure is required. The information system is like an intranet, in that it provides the backbone for efficient information transfer so that the whole system can function as it is designed.

2.2.4 Zachman Framework

The Zachman enterprise architecture framework was initially developed in the 1980s by John Zachman for the US government. Using consulting experience with large-scale IT systems development, the framework was extended to cover structures in other industry sectors. In the 1990s, the originator formed a consulting company and marketed the Zachman Framework to industry.

The Zachman Framework can be illustrated by a 6×6 matrix with themes called interrogatives running vertically and horizontally. The vertical theme is the communication interrogative, while the horizontal theme is called the reification transformations. Practically, each cell in the matrix represents a section of an enterprise focusing on the specific requirements relevant to the intersection of a pair of themes. According to Zachman, this structure is a generic representation of almost any system and can be regarded as an enterprise architecture framework due to its completeness. To assist discussions here, Table 2.1 shows a simplified version of the framework.

TABLE 2.1

Zachman Framework—A Simplified Illustration

	What	How	Where	Who	When	Why
Scope	What is important to business	Process the business performs	Location of business	Organization of business	Timing of business	Objectives of business
Enterprise model	Semantic model	Process model	Logistics model	Workflow model	Master schedule	Business plan
System model	Logical data model	Functional architecture	Distributed system architecture	Human organization	Process planning model	Business rule model
Technology model	Physical test model	System design	Technology architecture	Roles and responsibilities	Control structure	Rule design
Detailed representation	Database	Program	Network architecture	Security architecture	Timing control	Rule specification
Functioning enterprise	Data	Function	Network	People	Time	Motivation

The communication theme contains primitive interrogatives: what, how, when, who, where, and why. These are the six basic questions to clarify anything within a system. These themes apply to all cells in their respective columns. The vertical reification transformation theme contains concepts that need to be elaborated in the system: scope, enterprise, system, technology, detail, and function. As each of the communication themes is expanded into these concepts, the actual working mechanisms can be created and implemented to the full system.

The Zachman enterprise architecture is primarily a list of things to be created within the system. The advantage of this framework is its completeness. Using the Zachman Framework, the system design team has less chance of missing important features in the system's development.

2.2.5 DoDAF

The US Department of Defense Architecture Framework (DoDAF) is an overarching, comprehensive framework and conceptual model enabling the development of architectures to facilitate sharing of knowledge within the US Department of Defense (DoD). Managers in the DoD are required to specify the requirements and control the development of architectures for any defense systems they procure. Since it is the preferred system architecture, suppliers to the DoD should adopt DoDAF as the primary baseline for any defense systems they are contracted to build.

All modern defense systems are operated by complex systems with a Command, Control, Communications, Computers, Intelligence, Surveillance and Reconnaissance architecture framework (C4ISR AF, or simply C4ISR). The core of DoDAF is a data-centric approach where the creation of architectures to support decision-making is secondary to the collection, storage, and maintenance of data needed to make efficient and effective decisions. Figure 2.6 shows an overview of the conceptual framework of DoDAF/C4ISR.

The system architect and stakeholders of the system select views to ensure that architectures will explain current and future states of the process or activity under review. By selecting the appropriate architectural views, everyone in the team ensures that the requirement and proposed solution are explained and understood in the right ways. By systematically following the DoDAF process through all the blocks, i.e., the requirements, viewpoints, models, content, and visualizing in predefined ways, the system designer (often known as the enterprise architect) can design a system that is "fit-for-purpose" systematically.

To learn more about this AF, visit the DoDAF official website at http://dodcio.defense.gov/Library/DoD-Architecture-Framework/

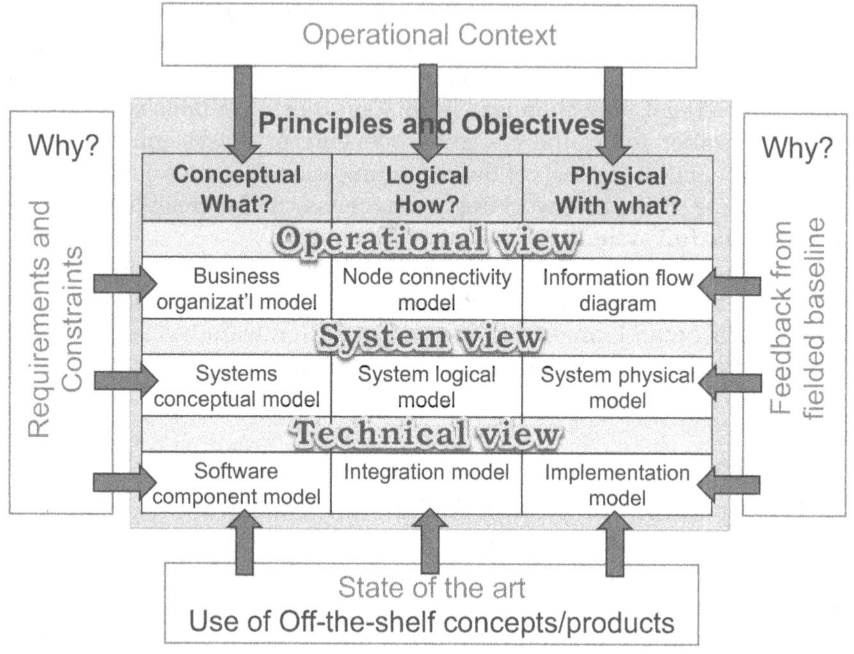

FIGURE 2.6
DoDAF/C4ISR concepts

2.2.6 TOGAF

Unlike country-specific groups such as the DoD, the Open Group is a global consortium that enables the achievement of business objectives through IT standards. The Open Group has more than 500 member organizations, mostly in the IT industry, and promotes the use of a standard system architecture to achieve these aims:

- Ensuring that everyone refers to same meaning for terms used in the system.
- Avoiding locking in to proprietary solutions by standardizing on open methods for enterprise architecture.
- Saving time and money, and utilize resources more effectively.
- Achieving a demonstrable return on investment.

The Open Group Architectural Framework (TOGAF) provides the methods and tools for assisting in the acceptance, production, use, and maintenance of an enterprise architecture. It is based on an iterative process model supported by best practices and a reusable set of existing architecture assets.

TOGAF has four architecture domains that are commonly accepted as subsets of an overall enterprise architecture:

- *Business Architecture* defines the business strategy, governance, organization, and key business processes.
- *Data Architecture* describes the structure of an organization's logical and physical data assets and data management resources.
- *Application Architecture* provides a blueprint for the individual applications to be deployed, their interactions, and their relationships to the core business processes of the organization.
- *Technology Architecture* describes the logical software and hardware capabilities that are required to support the deployment of business, data, and application services. This includes IT infrastructure, middleware, networks, communications, processing, and standards.

TOGAF specifies a series of architecture development activities carried out within an iterative cycle of continuous architecture definition and realization. This allows organizations to transform their enterprises in a controlled manner in response to business goals and opportunities. The cycle is illustrated in Figure 2.7.

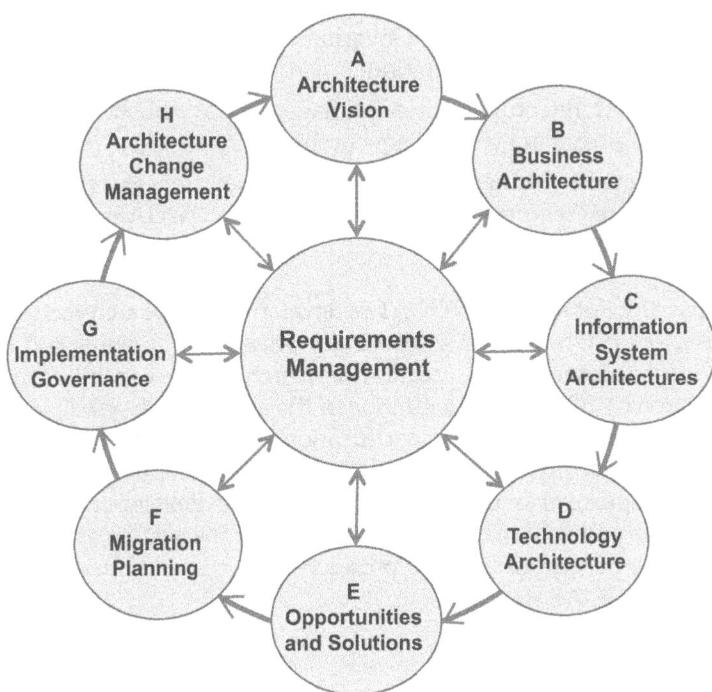

FIGURE 2.7
TOGAF processes

In Figure 2.7, each bubble represents a phase of development.

- **Phase A** (Architecture Vision) describes the initial phase of an architecture development cycle. It includes information that defines the scope of the architecture development initiative, identifying the stakeholders, creating the architecture vision, and obtaining approval to proceed with architecture development.
- **Phase B** (Business Architecture) is the development of a business structure to support the agreed architecture vision.
- **Phase C** (Information Systems Architectures) describes the development of an integrated information system to support the agreed architecture vision.
- **Phase D** (Technology Architecture) describes the development of technologies to support the agreed architecture vision.
- **Phase E** (Opportunities and Solutions) conducts initial implementation planning and the identification of delivery vehicles for the architecture defined in the previous phases.
- **Phase F** (Migration Planning) addresses how to move from the baseline to the target architectures by finalizing a detailed implementation and migration plan.
- **Phase G** (Implementation Governance) provides an architectural oversight of the implementation.
- **Phase H** (Architecture Change Management) establishes procedures for managing change to the new architecture.
- **Requirements Management** monitors the process of managing architecture requirements throughout the TOGAF architecture development methodology.

In preparation for applying TOGAF to develop a suitable architecture for the enterprise, TOGAF requires a **preliminary phase** that prepares and initiates the activities necessary for the creation of an architecture capability, including customization of TOGAF and definition of the architecture principles used.

In addition, TOGAF provides certification to systems that have architecture developed according to TOGAF enterprise development methodology. Certification approval is important to the system developer. It means the design reflects practical market requirements for the system as well providing a warranty of conformance for products, services, and business practices that meet the TOGAF standards.

2.2.7 MoDAF

The British Ministry of Defence Architecture Framework (MoDAF) is an internationally recognized enterprise architecture framework used by the

British MoD to support their project planning and change management activities. MoDAF provides managers with a comprehensive tool to aid in the understanding of the key factors they need to consider when making decisions. The MoD works closely with its international allies to ensure that when operating in coalition operations, capability information is shared to support interoperability. To facilitate this requirement, MoDAF was developed from DoDAF, but modified by MoD to include Strategic, Acquisition and Service Oriented viewpoints.

MoDAF specifies a set of rules and templates, known as Views, which provide a visualization of a particular business area within an enterprise, which are aimed at the various stakeholders who interact with the enterprise. These Views represent a window into the enterprise architecture that is specific to the particular stakeholder conducting the viewing. The Views are divided into seven categories:

- Strategic Views—These Views define the desired business outcomes and the capabilities required to achieve those outcomes in line with MoD's strategic intent.
- Operational Views—These Views define, in abstract, the processes, information, and entities needed to fulfill the capability requirements of the system either in a TO BE architecture or in an AS IS architecture.
- Service Oriented Views—These Views describe the services that are required to support the processes described in the Operational Views.
- Systems Views—These Views describe the physical implementation of the Operational and Service Orientated Views used to realize capability.
- Acquisition Views—These Views describe the dependencies and timelines of the projects that will deliver the solution, including dependencies between projects and capability integration across Defense Lines of Development.
- Technical Views—These Views are used to define the standards, rules, policies, and guidance that are to be applied to the architecture.
- All Views—These Views are used to provide a description and glossary of the complete project architecture, including scope, ownership, timeframe, and all other meta data that is necessary to effectively query architectural models.

The Views seen by the user of MoDAF are snapshots of the underlying structural data. The same information may be presented in more than one view and remains consistent across multiple views due to the reference model. Figure 2.8 illustrates how the different viewpoints relate to each other. The Strategic, Operational, and System viewpoints have a layered relationship;

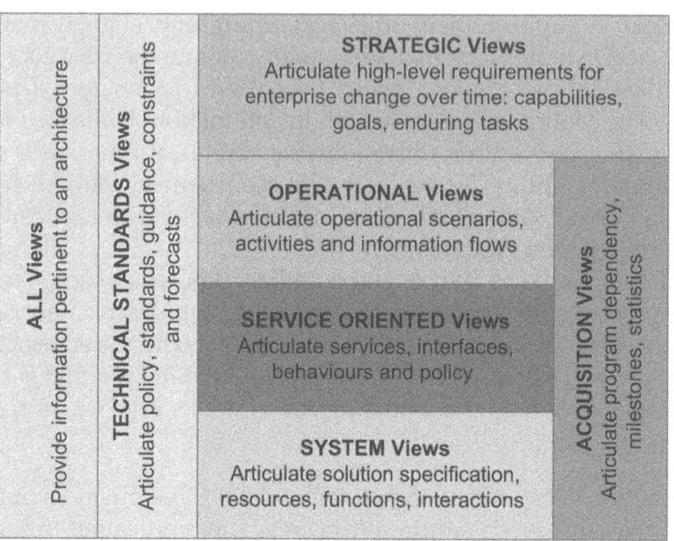

FIGURE 2.8
MoDAF viewpoint relationships

the Acquisition viewpoint sits beneath the Strategic viewpoint but has a supporting role across the Operational and System viewpoints. The All Views and Technical Standards Views stand alongside the other views as they provide information on supporting standards, ontology, and architecture description.

2.2.8 Popularity of Architectures

There is a long history of research into enterprise architectures by many organizations, not only in engineering, but also in other fields, including government, resources, IT, utilities, and defense. In 2003, the Institute for Enterprise Architecture Developments (IFEAD) carried out a survey to determine how these architectures have been used in different sectors. Table 2.2 shows the results. See http://www.enterprise-architecture.info/info.html for full details of the survey.

It should be noted that almost one-third of the organizations surveyed (32%) are using their own architecture framework (AF) to develop their internal and customer systems. An enterprise architecture is a comprehensive representation of the experience of many different business systems. Organizations that have good knowledge of their own business environment may already have considerable experience of the enterprise modeling process and may not be keen to implement a different AF from an external source. Full implementation of an enterprise architecture (EA) is time-consuming and requires a great deal of effort and resources.

TABLE 2.2

Survey Results of Enterprise Architecture Uses by IFEAD in 2003

Enterprise Architecture Framework	Acronym	Prevalence
Computer Integrated Manufacturing Open System Architecture	CIMOSA	6%
Purdue Enterprise Reference Architecture	PERA	3%
US Treasury Enterprise Architecture	TEAF	4%
ISO-IEC Guide to the POSIX Open System Environment	ISO/IEC 14252	3%
US Defense Architecture Framework	C4ISR	6%
Capgemini Integrated Architecture Framework	IAF	7%
The Open Group Architecture Framework	TOGAF	9%
US Federal Enterprise Architecture Framework	FEAF	6%
Zachman Framework		18%
Organization's own		32%
Other		6%

2.3 International Standards

The V-model described in chapter 1 is a generic representation of the steps involved in the development process of a system using a systems engineering approach. It provides little or no information about what the systems engineers and development team actually need to do. Systems development requires a breadth of knowledge and skills from the relevant branches of science and engineering, and as such cannot be simply intuited or learned from academic programs. Individual organization (*note*: systems engineering applications are not restricted to engineering companies) developing systems will accumulate experience over time, and will consolidate these experiences into standards in the form of written processes and templates for use in the future. These standards are the organization's own proprietary standards.

Different organizations therefore have developed different internal engineering processes due to their nature of their businesses and experiences. However, some business opportunities require that several organizations collaborate for their mutual benefit. In order to work with business partners, organizations need to synchronize their systems engineering processes to reduce conflicts and inconsistency.

In any industry, organizations realize the problems they face in partnering. Many industries have developed common knowledge, deliberately or by accident, which has become known as the de facto industry standards. Large corporations are often able to influence these de facto standards to include more of their internal processes so that the effort of adaptation to an agreed process when partnering is minimized. Smaller organizations also try to reduce the problems of adaptation by adopting most of the industry de facto standards as their standard operating procedures.

From the international perspective, standards are affected by increasing global trade. Organizations tendering for projects across national borders need suitable guidance on their offerings. The International Organization for Standardization was founded to promote and create international standards that aim to resolve conflicts that arise when organizations collaborate internationally. Many conflicts exist, including differences between internal processes of organizations, national practices and standards, cultural differences, infrastructure, etc. Such standards serve as the lowest common denominator for cross-company activities.

2.3.1 ISO 15704

In parallel with the efforts of individual organizations to develop systems engineering standards, the "enterprise architecture" industry (i.e., the Community of Common Interest in defining EAs) formed a task force in the 1990s to explore the best options for the standardization of EAs. The resulting International Standard ISO 15704 was developed to harmonize the differences between different architectures used in industrial automation systems.

The outcome of this standardization effort was to define two important aims of an enterprise reference architecture and related methodologies. These are:

- An enterprise reference architecture should:
  - Model the whole life history of an enterprise integration project from its initial concept through definition, functional design or specification, detailed design, physical implementation or construction, and operation, to decommissioning or obsolescence
  - Encompass the people, processes, and equipment involved in performing, managing, and controlling the enterprise mission.
- An enterprise reference architecture and related methodologies should:
  - Enable an enterprise integration planning team to determine and develop a course of action that is complete, accurate, properly oriented to future business developments, and carried out with the minimum of resources, personnel, and capital.

The standard does not exclude any existing EA, but rather points out that most EAs have pros and cons in their structure and methodologies. There is no one perfect EA and it is important that users of these architectures should be aware of the limitations when they develop their own systems.

2.3.2 GERAM

The Generalised Enterprise Reference Architecture and Methodology (GERAM) system was developed in parallel with ISO 15704 by another international group of systems engineers. GERAM provides a set of mapping concepts that can be used to explore the completeness of any existing EA. Figure 2.9 shows the framework of GERAM.

GERAM outlines the constituents of an EA and the relationships between them, to help determine whether an EA is able to fulfill the requirements of a system.

Starting at the top left-hand box, the Generalised Enterprise Reference Architecture (GERA) is the key part of any EA that employs an enterprise engineering methodology (EEM). This is a standard methodology (in the form of a set of procedures) to explain the content of the GERA. EEM in turn uses enterprise modeling languages (EMLs), mostly in the form of symbols, logics, or flow diagrams in its methodologies to visualize the GERA. The actual document containing Figure 2.9 and elaboration of the diagram is in fact an annex of ISO 15704.

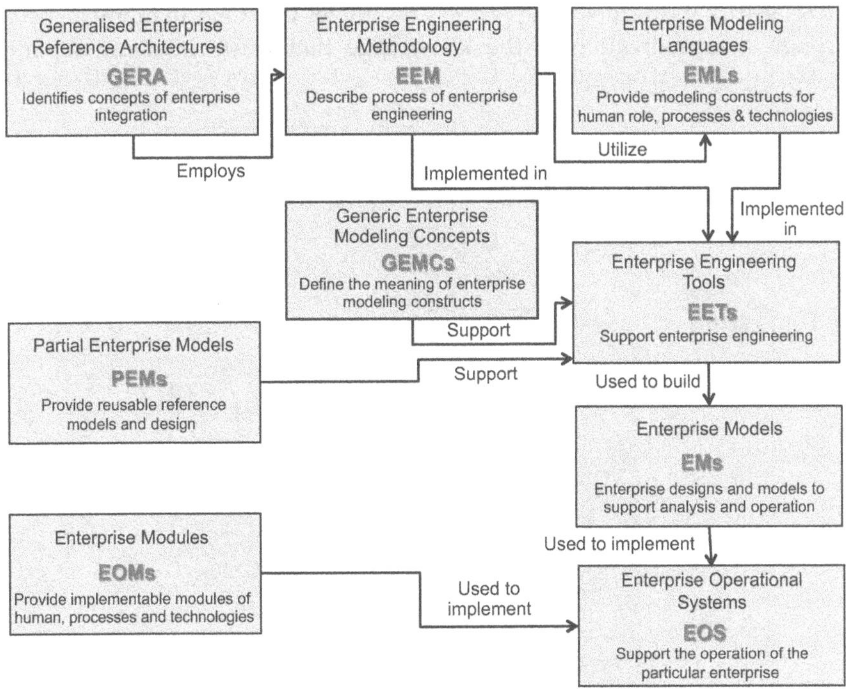

FIGURE 2.9
GERAM models

2.3.3 ISO 19439

In addition to ISO 15704, the ISO task group went on to develop ISO 19439, which specifies a framework for enterprise modeling and integration that conforms to ISO 15704.

- ISO 19439 defines and specifies the generic concepts that are required to enable the creation of enterprise models for industrial businesses and to provide support for the use of frameworks by industrial enterprises.
- This International Standard specifies that a framework for enterprise modeling should have certain features:
  - Both functional and informational views
  - Be able to derive resource and organization views
  - Provide distinct model phases (see Figure 2.10)
  - Provide for derivation of partial and particular models from generic constructs
  - Propagate model changes to all views

More interestingly, ISO 19439 describes an enterprise system modeling process (Figure 2.10).

It is clear that this modeling process resembles CIMOSA in many respects, through a formalization of the knowledge that exists in many separate

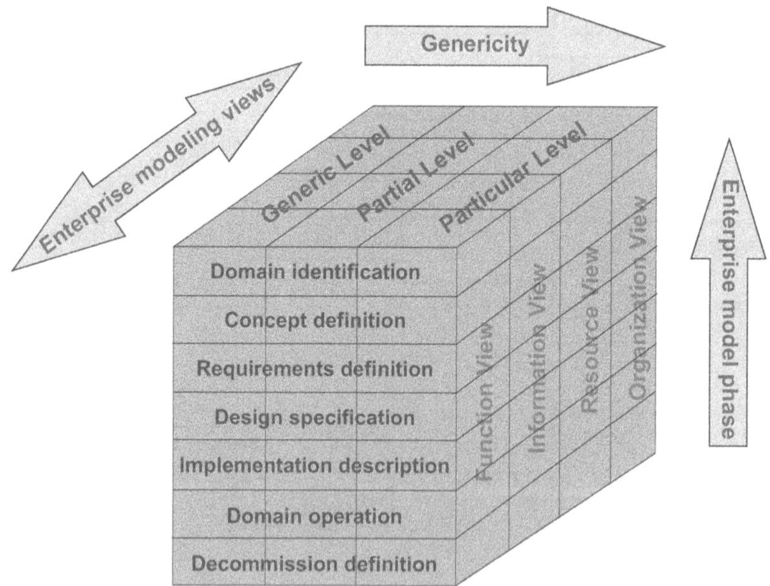

FIGURE 2.10
ISO 19439 modeling process

architectures. Using ISO 19439, developers of enterprise architectures are able to benchmark against any other EA on a standard comparison platform.

2.4 Case Study

This case study is the outcome of an investigation of the design of the through-life logistics support (TLS) system of a type of warship known as an Air Warfare Destroyer (AWD) [1].

The Australian Government and its defense organizations have struggled to effectively support defense capabilities throughout their operational life. Numerous reports have indicated management and structural failures that began during the procurement phase, with insufficient budget being apportioned to full lifecycle costs. These failures include not learning from previous projects, poor risk management practices, lack of responsibility and accountability, and inadequate communication amongst stakeholders. Government thinking has typically favoured the use of a performance-based contract known as a Contract for Availability (CA). However, research has shown that effective CfA contracts only work when adopting proven EA frameworks. This case study evaluated EA frameworks and suggested a suitable framework for establishing a TLS system for the Hobart Class AWD, covering its 30-year in-service life.

The qualitative comparison in Table 2.3 shows that there are pros and cons for each of the EAs short-listed for selection. Unless a "new" EA that

TABLE 2.3

Pros and Cons of Five Architecture Frameworks

	Pros	Cons
Zachman	• Externalize meanings of enterprise objects • Good as a guide for developing the EA	• Does not include method to create EA • No information on processes
TOGAF	• Open source, assessable by anyone • Lots of documentation, commentary, and user communications	• Heavily customized • Needs extensive training
FEA	• Used by US Federal agencies	• Primarily designed for procuring IT systems
DoDAF	• Use different views to explore the EA • Data visualized in graphics	• More complex EA structure
MoDAF	• Interoperable with DoDAF and other Architecture Frameworks used by British military allies	• More complex EA structure

combines all the pros and eliminates all the cons can be found, the AWD support organization should make a choice from one of these existing architectures. Hence, a multi-criteria decision analysis is required to select the most appropriate AF.

The outcome of a multi-criteria decision method depends on the correct understanding of the philosophy behind each of the methods. The weighted linear average method, which is probably the simplest and most widely used in industry, was chosen for this analysis for ease of communication with stakeholders. The essence of this method is the determination of a reasonable set of weights for the criteria. In this case, there was no literature to indicate the importance of particular criteria and hence all criteria carried equal weight.

An AF was scored against a range of criteria to determine its suitability based on the needs of the organization. The criteria were divided into five broad categories: (1) objectives; (2) properties; (3) components; (4) functions; and (5) services.

Each of the AFs selected in this study were rated against 26 criteria expanded from the five categories on a score sheet. This gives a visual representation of the strengths and weaknesses of the selected AFs against each of these criteria. From these results, either the best-fit candidate could be chosen, or if none of the candidates fulfills the criteria satisfactorily then a hybrid solution can be selected based on the scores.

The process of scoring the short-listed AFs using the suggested methodology against the organization's criteria can be subjective. Ideally, the group of stakeholders who helped to develop the organization's criteria would be involved in the AF selection process. This same group of stakeholders could then score each of the short-listed AFs according to the criteria. In this case study, a small group of five integrated logistics and management system practitioners with wide-ranging experience in developing and supporting defense projects was gathered to workshop the criteria questions. A series of meetings were held over a number of weeks to inform the team members of the reasons behind the need to select an AF, the purpose and benefits to be gained from using an AF, details of the short-listed AFs, and to examine the criteria. The purpose of these meetings was to ensure that each of the team members fully understood what questions were being asked of the AF being examined, and had sufficient background and detail to be able to score the AF against the criteria.

Once the individual tasks were completed, the team then met to discuss their individual scores, and explain the reasoning behind any scores that differed wildly from the group average. The results are recorded in Table 2.4.

From the summary score sheet, the best choice of AF for the Hobart Class AWD is MoDAF. The list of AFs short-listed shows a progression through the history of AF. Zachman, who first posited the idea of a framework for

TABLE 2.4

Summary of Score Sheet for Different AFs

	Methodology				
	Zachman	**TOGAF**	**FEA**	**DoDAF**	**MoDAF**
1 Guide	2	5	6	7	8
2 Centralize	3	5	6	7	7
3 Simplify	3	5	6	7	9
4 Standardize	2	4	6	8	9
Objectives total	**10**	**19**	**24**	**29**	**33**
5 Customizability	2	4	6	7	8
6 Correctness	1	3	4	7	7
7 Compatibility	2	4	4	8	8
8 Completeness	2	4	5	5	9
9 Conciseness	2	3	5	6	7
10 Subsetting	1	3	7	7	8
Properties total	**10**	**21**	**31**	**40**	**47**
11 Deliverables	3	5	6	7	9
12 Methods	2	4	5	8	8
13 Techniques	1	2	5	7	8
14 Standards	6	6	6	7	8
15 Roles	2	3	5	6	8
16 Tools	2	2	4	7	7
17 Paths	2	3	4	6	8
Components total	**18**	**25**	**35**	**48**	**52**
18 Management	2	4	5	7	9
19 Navigation	6	7	7	7	7
20 Handbook	1	2	4	7	7
21 Interfaces	3	5	5	7	8
Functions total	**12**	**18**	**21**	**28**	**31**
22 Education	2	2	2	8	8
23 Integration	3	3	2	7	8
24 Demonstration	1	1	2	6	8
25 Customization	1	3	4	7	7
26 Implementation	5	3	4	7	9
Services total	**12**	**12**	**14**	**35**	**40**

organizing the documentation and processes of an organization, scores lowest. It may appear that the selection of MoDAF was based on the assumption that as the newest of the AF examined, it was regarded as the best suited. This is not strictly true, as each of the AFs examined had deficiencies which counted against them when they were scored against the assessment criteria.

2.5 Reflections

Using your experience of working on systems development projects, consider these questions.

- What other objectives you think should be considered in addition to the 26 objectives described in the paper? Nominate and explain at least one objective from your experience.
- If, for some reason, you have to exclude MoDAF and DoDAF from your selection list, nominate two other enterprise architectures in their place and carry out your own analysis to select the most suitable EA for this case. You need to include descriptions of your previous experience in the above nominated enterprise architectures. Justify your selection with ratings and rationale for those ratings.

The selection method used in the case study is one possible way to choose an EA framework. However, the outcome of this method can vary if different groups of people are involved. Assuming that you have a large-scale system development project to be commissioned (different from the one described in the paper), consider your case with the following focus questions:

- What does "solution architecture" mean to you? How can you describe the required solution architecture?
- Who are the true stakeholders in the solution architecture of your large-scale system?
- Are there any outside influences that will affect the design and selection of the solution architecture?
- What other selection methods do you suggest to help you to select the solution architecture?

2.6 Reference

1. Thompson, D. & Mo, J.P.T. (2015). "Optimum Support System Architecture for Air Warfare Destroyers." In *Transdisciplinary Lifecycle Analysis of Systems*. Stjepandić, J. & Curran, R. (Eds.), 22nd ISPE Concurrent Engineering Conference, 20–23 July, 2015. IOS Press.

2.7 Additional Reading

Bernus, P., Nemes, L. & Schmidt, G.J. (Eds.) *Handbook on Enterprise Architecture*. Springer, 2003. ISBN: 978-3-540-24744-9, DOI: 10.1007/978-3-540-24744-9

3

System of Systems Framework and Environment

> The ideal architect should be a man of letters, a skillful draftsman, a mathematician, familiar with historical studies, a diligent student of philosophy, acquainted with music, not ignorant of medicine, learned in the responses of jurisconsults, familiar with astronomy and astronomical calculations.
>
> **Marcus Vitruvius Pollio**
> *Military engineer, first century BCE*

3.1 Characterizing a System of Systems Framework and Environment

It has been observed that complex systems need to be considered from multiple viewpoints, or put another way, we need to assemble "blocks" of complementary data that inform system design and operation. The system of systems design context is characterized by the consideration of "scale, complexity, and certain non-technical constraints necessitating the use of architecture methods and approaches that are different from those used for system architectures" [1]. From an automotive perspective, Pelliccione et al [2] have observed that

> During the past 20 years vehicles have become more and more robot like, interpreting and exploiting input from various sensors to make decisions and finally commit actions that were previously made by humans. Such features will require continuous evolution and updates to ensure safety, security, and suitability for supporting drivers in an ever changing world. Modern vehicles can have over 100 Electronic Control Units (ECUs), which are small computers, together executing gigabytes of software. ECUs are connected to each other through several networks within the car, and in some cases also to the outside world. This need for addressing ever increasing complexity as well as for offering flexibility, support of continuous evolution, and very late changes in user visible features introduces new challenges for developing and maintaining a suitable electronic architecture.

3.1.1 Introducing ISO 42010

In this chapter, we introduce such a framework by adapting the ISO standard 42010, titled "Systems and software engineering—Recommended practice for architectural description of software-intensive systems". It is stated in the standard that *"this recommended practice addresses the activities of the creation, analysis, and sustainment of architectures of software-intensive systems, and the recording of such architectures in terms of architectural descriptions"*. The standard was developed over more than a decade with inputs from many experts from different countries and industries, with the following intentions:

- To define useful terms, principles, and guidelines for the consistent application of architectural precepts to systems throughout their lifecycle.
- To elaborate architectural precepts and their anticipated benefits for software products, systems, and aggregated systems (i.e., "systems of systems").
- To provide a framework for the collection and consideration of architectural attributes and related information for use in other standards.
- To provide a useful road map for the incorporation of architectural precepts in the generation, revision, and application of other standards.

It is noted that this framework may be adopted to both characterize a large, complex system and to characterize components of such systems.

A diagram illustrating the elemental parts of the standard is shown in Figure 3.1.

The operating environment frames conditions at the boundary of a particular system of systems. As systems-of-systems thinking becomes more widespread, the term ecosystem is being used to characterize the environment where a particular application is embedded: the environment beyond the boundary. Borrowed from the life sciences, where it means a biological community of interacting organisms and their physical environment, this term is now being applied to complex networks or interconnected systems. Some authors refer to software ecosystems. Some refer to manufacturing ecosystems, business ecosystems, and entrepreneurial ecosystems, or to cultural ecosystems within an enterprise. We shall discuss particular instances later, but there are some common themes:

- A particular system of systems may itself be viewed as a system within a broader system, and interactions between the two may have unexpected consequences.
- Ecosystems are comprised of semi-autonomous agents and evolving support platforms where changes in one can lead to unexpected effects on another.

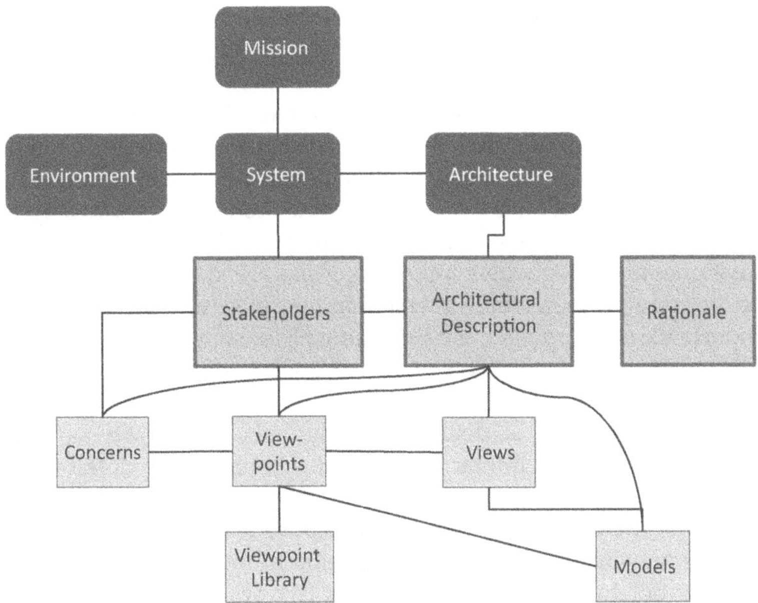

FIGURE 3.1
A representation of ISO 42010 elements and some linkages between them

- There is an orientation towards finding configurations that are dynamically balanced and sustainable at a particular point in time. If the system or the environment changes, new connections may have to be established.

The smartphone provides an example. It has internal systems that may be interconnected (e.g., to take a photo, store it and share it) and connections with external systems (e.g., a telephone network, a community of friends, an app store) and the owner can configure the phone to suit his or her preferences. In this book we are concerned with operational systems (e.g., the smartphone) where an asset is supported to sustain its operation (e.g., via the manufacturer's software updates), and we need to recognize that both the asset and its support system interact through larger systems (e.g., the internet) that are not controlled by either. The proliferation of spam and fake news provides an example of unintended effects in this system of systems.

3.1.2 Introducing Matrix Thinking

The ISO 42010 representation of a system and its operating environment shows some interactions between the elements (Figure 3.1). In Figure 3.1, boxes represent classes of things. Lines connecting boxes represent associations between things. An association has two roles (one in each direction).

Within the body of the standard there is reference to other interactions, e.g., the rationale shall address the extent to which the stakeholders and concerns are covered by the viewpoints selected. Our experience in using this framework has highlighted the existence of many more interactions, as would be expected in a systems-of-systems scenario. We have adapted the interaction matrix concept from systems engineering to consider such attributes.

Simple 2 × 2 matrices have been used to map scenarios for a variety of purposes (for some examples see [8]). More complex matrices have been utilized in computational modeling where interaction rules are associated with each connection, or to simply map where interactions take place and highlight their relative importance (see Eppinger and Browning [9] for some general background).

An example of an interaction matrix showing connections between elements of the ISO 42010 framework is shown in Table 3.1 for discussion purposes. At this level of detail, the diagonals are shown as null (e.g., stakeholder–stakeholder interaction), but at a finer level of detail there may be independent interactions (e.g., between different stakeholders). The example highlights the connections shown in Table 3.1, which suggest that some interactions may be one-way (e.g., the mission influences the system, but the system may not influence the mission) or two-way (e.g., there is an interplay between the system and the architectural description). In many places throughout the ISO 42010 document there are references to interactions

TABLE 3.1

An ISO 42010 Element Interaction Matrix

		To											
		Stakeholders	**Architectural Description**	**Rationale**	**Mission**	**Environment**	**System**	**Architecture**	**Concerns**	**Viewpoint**	**Viewpoint Library**	**Views**	**Models**
From	Stakeholders	X	I	I			I		I	I			
	Architectural description	I	X				I					I	I
	Rationale		I	X									
	Mission				X		I						
	Environment					X	I						
	System	I	I		I	I	X	I					
	Architecture						I	X					
	Concerns		I						X	I			
	Viewpoint	I	I							X	I	I	I
	Viewpoint library									I	X		
	Views		I							I		X	I
	Models		I							I		I	X

that correlate a row element with column elements. These are presented as "viewpoints". Examples from annex D of the standard are a link between mission and stakeholders, outlining system purpose, scope, and a link between mission and system, outlining roles played by the system.

3.2 Stakeholders, Architecture Description, and Rationale

3.2.1 Stakeholders

The ISO 42010 standard states that the principal class of users for this recommended practice comprises stakeholders in system development and evolution, including the following:

- Those that use, own, and acquire the system (users, operators, and acquirers, or clients).
- Those that develop, describe, and document architectures (architects).
- Those that develop, deliver, and maintain the system (architects, designers, programmers, maintainers, testers, domain engineers, quality assurance staff, configuration management staff, suppliers, and project managers or developers).
- Those who oversee and evaluate systems and their development (chief information officers, auditors, and independent assessors).
- A secondary class of users of this recommended practice comprises those involved in the enterprise-wide infrastructure activities that span multiple system developments, including methodologists, process and process-improvement engineers, researchers, producers of standards, tool builders, and trainers.

This implies that stakeholders may be classified in terms of the role or assignment they are responsible for, such as those mentioned above plus others we have observed: technology providers, dealers, researchers, regulators, customers, field officers, change agents, functional specialists, suppliers, and competitors. Some roles relate to an engineering phase and some to an operational phase. We specifically include competitors as they can influence the selected architecture.

In the previous chapter, we discussed the concept of enterprise architectures, and provided examples of different models for describing engineering and operational scenarios at an enterprise or project level. Adopting the position that enterprises and projects exist in a broader context, we introduce the notion that stakeholders may also be classified as entity types such as individuals, project or functional groups, enterprises, local communities, broader communities, special interest groups, regulatory authorities, and governments, each with associated roles. Who does what in designing a system of systems and who does what in operating a system of systems?

In our work with dynamically changing organizations, we have found that the concept of a responsibility matrix that maps generic roles against generic entities such as an individual or group provides a stable platform for assigning tasks. The concept has been used in coordinating the management of multiple projects that make different demands on enterprise functional departments at different stages of development, and in clarifying the roles and expectations of stakeholders in maintaining ethical relationships in the construction industry. Craig and Boley [3] describe the use of personal agents and rule-responder architecture where task assignments of individuals are defined in a responsibility matrix and the roles assigned are subsequently defined by an associated OWL (Web Ontology Language) Lite ontology. The point being made here is that this construct can be used in a variety of settings.

3.2.2 Architecture Description

An architecture description brings together the logic behind a particular system of systems at a conceptual level. As noted in the previous chapter, an architecture model has some typical characteristics:

- It captures previous experience in a knowledge structure.
- It provides a baseline to build new structure or content.
- It represents the minimum expectations.

The reference models presented and the practices adopted to coordinate complex projects may lead to document-centric architecture descriptions (e.g., DoDAF). However Bayer et al [4], in describing the complex interactions to be dealt with in the design and deployment of space missions, have explored a model-centric approach which supports the testing of different system perspectives and uses or re-uses.

ISO 42010 defines an architecture description as "a collection of products to document an architecture." Multiple viewpoints can be represented, and any inconsistency between viewpoints noted. It is suggested the rationale for the selection of architectural concepts should be represented, along with evidence of the consideration of alternative architectural concepts. When describing an adaptation of a pre-existing system, the rationale for the legacy system architecture should also be presented.

3.2.3 Rationale

ISO 42010 requires that system architects provide evidence of the consideration of alternative architectural concepts, and the rationale for considering these alternatives. It is also suggested the rationale for the selection of particular viewpoints utilized be enunciated. Documents related to rationale are seen as part of the architectural description. However, Emery and Hilliard [5] have observed that the standard gives little normative guidance for capture or to represent this information. We suggest that two sets of questions be asked to draw out such information.

- Set 1—we suggest a critical questioning approach using nine questions related to the system description rationale (see Figure 3.1). How has the architecture description been influenced by the following: stakeholders, mission, the environment, system functional requirement the choice of architecture, concerns, previous viewpoints, models, the selected viewpoint, and alternative views?
- Set 2—we take a knowledge-oriented approach: What background knowledge has influenced the choices made (see later discussion and Figure 3.2)?

Bosch [6] observed that the first phase of software architecture research, where the key concepts were components and connectors, has matured the technology to a level where industry adoption is wide-spread and few fundamental issues remain. But the lack of first-class representation of design decisions and the fact that these design decisions are cross-cutting and intertwined, and that this lack of representation leads to high maintenance costs. As a result, design rules and constraints are easily violated and obsolete design decisions are not removed as the system evolves.

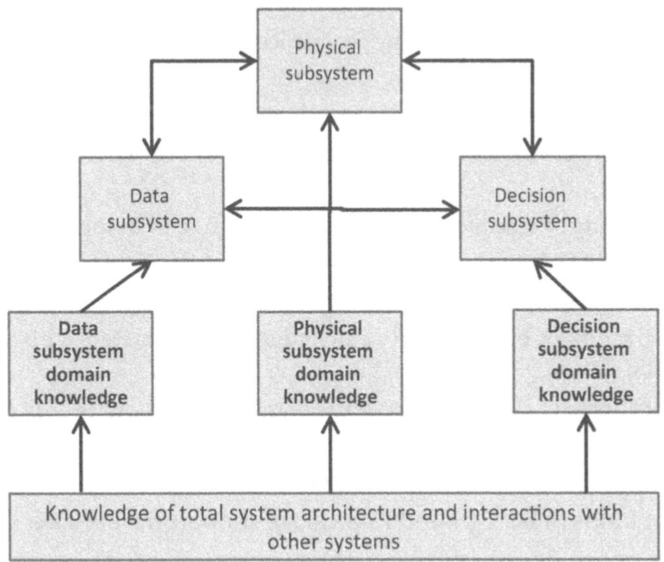

FIGURE 3.2
Knowledge linked in total system architecture

We have directly observed the impact of this kind of concern in relation to a supply chain management system that was part of an integrated enterprise resource planning (ERP) system. An effective system had evolved due to the joint efforts of users and developers. But when the users changed due to a major reorganization, the system began to fail at times. The IT experts decided the problem was at the human-computer interface, and introduced updates. This had minimal impact. It was finally recognized that the IT system was only part of the total system, with the other part being knowledge tacitly held by the previous users, and the rationale for the particular system configuration included assumptions about human action and decision-making. This is illustrated in Figure 3.2.

The rationale for a particular architecture might be driven by a need for change, or compliance with standards that may also stimulate innovation, or a need for adaptability. A modern military aircraft provides an example of all three being combined in a loosely coupled systems architecture where the physical, information, and decision system elements are linked via a data-bus protocol and its related management system. Subsystem elements can be readily changed without disturbing the overall system as long as the data exchange protocols are honored. The rationale forms part of the system knowledge base, and the rationale adopted is informed by domain knowledge of physical, information, and decision subsystems, and architectural knowledge about how they work together and link with other systems.

This perspective leads to the second set of questions (Set 2) that might be asked:

- What are the activities or physical processes at work? Who does what to make them happen?
- What information is needed to undertake each physical process?
- What background knowledge is needed to support each physical process?
- What kinds of technical data have to be managed in the data system?
- What kinds of management and control data have to be managed in the data system?
- What kinds of background knowledge are associated with managing the data?
- What kinds of decisions have to be made, and how is this process supported?
- What kinds of data are needed to make decisions?
- What kinds of background knowledge support the decision system?
- What overarching architectural knowledge is needed to integrate the physical/data/decision systems and understand linkages between these and other functional systems?

3.3 Mission and Functionality

3.3.1 Mission

Our objective is to develop, deploy, operate, or sustain an asset viewed as a system of systems. The ISO 42010 standard states that "a system exists to fulfill one or more missions in its environment. A mission is a use or operation for which a system is intended by one or more stakeholders to meet some set of objectives."

A variety of asset types may be linked in networks: financial assets, physical assets, intangible assets, and knowledge assets to achieve the mission. For example, we may consider financial networks that are facilitated by the application of knowledge, intangible (data, software), and physical assets in smart banking systems. In this context, ISO 19439 describes an enterprise system modeling process that starts with a domain identification, creating a link with the system environment. The TOGAF enterprise model described in chapter 2 has four architecture domains that are commonly associated with specific mission objectives:

- Business Architecture defines the business strategy, governance, organization, and key business processes supporting economic sustainability.
- Data Architecture describes the structure of an organization's logical and physical data assets and data management resources supporting decision-making.
- Application Architecture provides a blueprint for the individual applications to be deployed, their interactions, and their relationships to the core business processes of the organization.
- Technology Architecture describes the logical software and hardware capabilities that are required to support the deployment of business, data, and application services. This includes IT infrastructure, middleware, networks, communications, processing, standards, etc.

3.3.2 System Functionality

In thinking about the systems element of the ISO 42010 framework from a system-of-systems perspective, it is first necessary to declare the boundaries: what functionality is required within the system to achieve the desired mission and what linkages with the broader ecosystem (e.g., internet connectivity) are needed to make it work. In developing an architectural framework, we have commonly focused on identifying functionality to complete all aspects and phases of the mission independent of the means of implementation.

In our application of the ISO 42010 framework, a system is viewed as a set of interlinked functional activities that may be represented at varying levels of detail. We have most commonly used a relatively simple modeling tool [10] to represent each functional activity within a system-of-systems component. At each successively finer level of detail, the same questions are asked: what inputs are needed to facilitate this activity, what outcomes are sought, what are the "rules for the game", and what kinds of resources are needed. Finally, we ask how the activities are connected.

3.3.3 System Architecture

Creating an architecture is concerned with the development of functional and achievable system concepts, maintaining them through development, certifying built systems for use, and verifying these concepts during operational and subsequent evolutionary phases. Detailed systems engineering activities, such as comprehensive definition of requirements, interface specification, and defining the architecture of major subsystems are tasks that typically follow on from development of the system architecture (ISO 32010).

The activity may relate to the development of an individual system, the iterative development of evolving systems, the formalization of the architecture of existing systems, or the evaluation of the architecture of an existing system. In the previous chapter, we briefly described the practice of transition planning (Figure 2.2)—describe an AS IS architecture, and a TO BE architecture, developing plans to fill the gap and researching missing technology. This may also include acquiring missing knowledge.

In chapter 2, we proposed that the architecting of a building requires consideration of stakeholder, functional, and technical requirements, plus recognition of boundary conditions, such as physical access, access to power and water, and the associated ecological environment. These same considerations apply to, architecting a system of systems. The systems view informs us what has to be done. The architecture view advises how it is to be done. We have to decide what is to be done by people or agents and what is to be done by technology? Are linkages between the system components and subcomponents to be fixed, or are linkages and the utilization of particular components reconfigured for different missions? What resources are to be developed or drawn on? How the boundaries are defined—what is to be done within the core system and what is to be done outside of it? We begin to look at complex systems of systems from multiple viewpoints.

3.4 Drivers and Multiple Viewpoints

The architecture description of systems commonly reflects a need or an opportunity to innovate combined with the prior experience of a number of stakeholders. This is reflected in the ISO 42010 framework under the headings of concerns, viewpoints, and models. In the following we have supplemented the framework content with some of our own experience of using it.

3.4.1 Concerns

In Table 3.1, the row relating to concerns shows that concerns may arise from many different sources such as the ability of the system to achieve its mission, system development feasibility, specific concerns of stakeholders. The ISO 42010 documentation also associates concerns with viewpoints:

- The structural viewpoint—What are the computational elements of the system and how are they organized? What are the software elements of the system, what interfaces and connections are needed, and what are the interconnection mechanisms?
- The enterprise viewpoint—What role is played by the system in achieving the mission and what activities does it support? Is the system consistent with the enterprise policy?
- The information viewpoint—What are the semantics of information and associated processing, and how is the language defined?
- The engineering viewpoint—What node structure, connection channels, and interfaces need to be considered, along with potential failure modes?
- The technology viewpoint—What captures the choice of technology specification and the adoption decision? How are specifications implemented, and what support for testing is required?

These kinds of concerns are most commonly associated with system development and operation. While using this standard, we have also reflected on concerns about the changing technological and operational environment in the broader ecosystem. What external momentum shifts (or what we may describe as "grand challenges") might lead to a need to redevelop or reconfigure an established system?

3.4.2 Viewpoints and Viewpoint Library

In the recommended ISO 42010 practice, the term viewpoint is used to designate a means for constructing a view that is independent of any particular system. The term was chosen to align with that of the ISO Reference Model of Open Distributed Processing (RM-ODP), which defines a viewpoint on a system as a form of abstraction achieved using a selected set of architectural constructs and structuring rules, in order to focus on particular concerns within a system. RM-ODP has no separate term for view.

The relationship between viewpoint and view is seen as analogous to that of a template and an instance of that template; or to use an object-oriented programming metaphor—viewpoint : view :: class : object. A viewpoint must be specified by the stakeholders, reflecting the concerns that it addresses and the languages, models, and techniques it employs. This provides a means to capture the considerable commonality found amongst existing architectures and techniques.

In view of their potentially broad application, it is recommended that viewpoints be documented to include information about their sources and assigned unique identifiers. A viewpoint specification may include additional information about the architectural practices associated with using the viewpoint:

- Formal or informal consistency and completeness tests to be applied to the models making up an associated view.
- Evaluation or analysis techniques to be applied to the models.
- Heuristics, patterns, or other guidelines to assist in synthesis of an associated view.

Multiple viewpoints, e.g., reflecting the experience of different stakeholders, may be assembled over time and stored in a library so that they can be reused as a form of template.

3.4.3 Models and Views

The ISO 42010 standard offers little advice in relation to models. As observed in chapter 2, the use of a particular enterprise model to inform the architectural description may be mandated in some situations (e.g., military procurement). In general, however, Table 2.2 shows that most organizations use their own model, and the next most popular model chosen is the Zachman Framework.

From our observations, an organization's own model reflects that organization's prior experience and may be most appropriate when adapting an as-is situation to enhance operations rather than when developing systems to support a different mission. Many of the models described in chapter 2 provide a structural view, whereas the Zachman Framework also asks questions about "who"—introducing the human role in systems operations, and "when"—introducing matters of timing. The TOGAF model (Figure 2.7) presents a process view, placing requirements management at the core in the context of phased development. We suggest this kind of process view may also be appropriate for systems of systems that are dynamically reconfigured to meet emergent needs at particular points in time, exploring configurations appropriate to different mission scenarios.

The ISO 42010 standard suggests that a view is a representation or description of the entire system from a single perspective. In contrast to a viewpoint, a view refers to a particular architecture of a system (i.e., an individual system, a product line, a system of systems, etc.) A view is primarily composed of models, although it also has additional attributes. The models provide the specific description, or content, of an architecture. For example, a structural view might consist of a set of models of the system structure. The elements of such models might include identifiable system components and their interfaces, and interconnections between those components.

3.5 Operating Environment Considerations

The operating environment may be considered from multiple perspectives: engineering and operational considerations, combined with the digital and application environment (Table 3.2).

The environment may on the one hand provide supporting infrastructure, e.g., high-speed internet, but on the other hand give rise to concerns by particular stakeholders, e.g., technological maturity.

In our studies, we have used ideas from the strategic management field in considering the emergence of "grand challenges" or opportunities to introduce change by reviewing current issues and future trends from following six perspectives:

- **Political perspective**—trends in government policy and regulatory interventions that may positively or negatively impact the configuration and economic viability of a particular system architecture.
- **Economic perspective**—what people are prepared and able to pay for and market trends.

TABLE 3.2

Environments in Multiple Perspectives

	Engineering Environment Considerations	Operational Environment Considerations
Digital and technology environment	• Development resources • Technology access • Technology maturity	• Reliability and adaptability • Access to supporting infrastructure
Application and stakeholder environment	• Mission/functionality • Temporal considerations • Competitive performance	• Mission sustainment • Dynamic capabilities • Access to supporting infrastructure • The broader ecosystem

- **Social perspective**—what social factors might favor a particular system or architecture.
- **Technological perspective**—what new technologies or technological infrastructure might be emerging and what has become obsolete.
- **Legal perspective**—are there any contractual or legal constraints that influence the adoption of a particular architecture.
- **Environmental perspective**—Are there ecological considerations, e.g. the demand to reduce power consumption that will influence the architecture adopted.

3.6 Case Study

Agricultural practices and supporting technologies have been co-evolving for decades, leading to dramatic improvements in farm productivity and establishing food production capabilities in places where this has not previously been seen as practical. An example is the development of hothouse technology that supports rapid growth of hydroponic vegetables in hot, arid areas near the sea by using the heat of the sun to desalinate seawater, or the establishment of semi-transparent hothouses for food production that also generate electricity.

The emergence of cyber-physical systems has seen the introduction of self-driving farm machinery guided by global positioning satellite data, the use of air, water, and soil sensors to help assess and manage growing conditions, and the use of drones to rapidly survey large farms for areas that may need special attention [7].

The farm-machinery manufacturer, John Deere, offers a service called FarmSight to complement their conventional equipment line. Smart sensors on board their equipment give early warnings when maintenance is required, and the company is working with specialist suppliers of other kinds of sensors to help farmers make timely decisions, e.g., about when soil conditions are right to plant seed. The company is also working with partners to develop business modeling tools to support financially sustainable farming operations.

Representing this scenario in terms of the ISO 42010 elements shown in Figure 3.1, we would suggest that the cyber-physical system **mission** is to enhance farm productivity, in combination with economic and ecological sustainability, e.g., using sensors to minimize water consumption or crop spraying requirements. The system supports the management of key events in time, e.g., ploughing a field to be ready for crop planting. The **system** functionality is customized for each farm or region, but includes both operational and support functions. Examples of the operational functions include a combination of physical, information, and decision systems. The physical systems may include autonomous machines, smart sensors, communications technologies, and a command and control center. The support functions may include training, maintenance, and logistical support for consumables and spare parts.

The system **environment** is framed in terms of the natural environment, e.g., location, weather, terrain, sunshine and water availability; in terms of social and economic norms and by accessible technological infrastructure with legal or contractual considerations. For example, is internet access available, will working and contractual arrangements create new interdependencies. The political environment may influence economics, e.g., if there is a government initiative to support the improvement of farm productivity. The system **architecture** defines what is done by people, what is done by technology, and how they are connected. The FarmSight initiative can support different combinations. Whatever the choice, there is a succession of events to be managed, leading to different configurations of subsystems being activated at different times, i.e., subsystem states may vary from dormant to active, whether associated with a particular event or continuously operating. The farmer's plough is only used at certain times, and maintenance may be required at regular intervals, but sensors for soil moisture content may be operating continuously. This suggests that discrete event analysis may support the development of a suitable system-of-systems architecture.

There is a multiplicity of **stakeholders,** depending on where the system boundaries are drawn within a larger ecosystem. There is the particular farm and its clients, and its local John Deere agent and IT partners. Each of these have connections into supply chains and communities of practice, and both the agent and the farmer are members of a local community where their reputations can be important. The appropriate **architectural description** will have different elements for different stakeholders. For the farmer, the description starts with the manufacturer's sales pitch:

> John Deere FarmSight gives you deeper insights into your business that were simply not possible until now. It enables individual remote monitoring of machines to ensure preventative maintenance, optimization for fuel consumption and output performance, as well as giving you new, more detailed agronomic information for better decision-making. John Deere FarmSight is available through a series of four service packages from your local John Deere dealer.

For the dealer, the description will outline how services are provided to multiple farms and how strategic technology partners and other suppliers are involved. The **rationale** for the establishment of the John Deere FarmSight program and engagement with it has to make business and operational sense for the farmer, considering the broader environment and any specific concerns. What makes business sense—will productivity be enhanced so the farmer can produce more from a given set of physical and financial assets, or can the total cost of a current operation be reduced sufficiently to offset technology acquisition and operational costs? Can the skills needed to operate in this way be easily acquired?

For the farmer, there are some **concerns** about the need to improve productivity that are driven by the external environment, and some concerns about the costs and reliability of the smart technologies being introduced. In particular, for smaller farms, the systems have to operate efficiently in the background as the farmer may not have time to attend to defective equipment at busy times in the farming calendar. For the John Deere agent, there may be concerns about their ability to support IT systems in the field, the economic viability of the service, and the preservation of their reputation as a reliable supplier and responsible community member. All stakeholders will draw on **views** and **viewpoint libraries** reflecting their own experience and those in their extended social networks. IT providers will draw on experience gained from other customers, and from their issue management logs.

In relation to **models,** the farming system may be viewed as a process with characteristic recurring stages, e.g., preparing for planting, planting seed, harvesting crops, etc. Each stage has a lifecycle, as represented in some enterprise models (see Chapter 2), with concepts and requirements remaining stable, but with potential variation in implementation arrangements depending on information from sensor systems, e.g., relating to soil moisture and weather conditions. Thinking about the processes of farming in this way can help provide a common understanding between system developers and system users in this environment.

3.7 Reflections

The core message here is that to be able to understand and develop complex systems of systems, we need not only to apply systems thinking at multiple levels of detail, but also to rationalize multiple viewpoints and apply network thinking. The need for multiple viewpoints is illustrated by the Indian parable about some blind men and an elephant. A group of blind men had never come across an elephant before, but learned about it by touching it, some from the sides, some from the back, and some from the front. Each individual had his own interpretation of what an elephant was, but as a group there was confusion until they could find a way to combine and rationalize their observations.

In this chapter, we introduced some aspects of complex systems design and operation to be considered, and we proposed matrix thinking as a means of looking at connections. Consider these questions based on your own experience:

- Can you identify a complex system where the viewpoints of multiple stakeholders had to be rationalized?
- How could you have brought these viewpoints together as an architectural description?
- How would you map interactions between organizational functions or within systems of systems familiar to you?
- What does each element of the ISO 42010 framework mean to you?
- Is the need for system development or adaptation stimulated by undertaking a new mission (e.g., putting a man on Mars), by concerns about what could be done differently (e.g., drawing on new tool), or opportunities from challenges emerging in the broader environment (e.g., a new technological infrastructure)?

3.8 References

1. Klein, J. & Van Vliet, H. (2013). "A systematic review of system-of-systems architecture research." In *Proceedings of the 9th international ACM Sigsoft conference on quality of software architectures*, pp. 13–22. ACM Digital Library - https://dl.acm.org/citation.cfm?id=2465478&picked=prox.
2. Pelliccione, P., Knauss, E., Heldal, R., Mallozzi, P., Alminger, A. & Borgentun, D. (2016). "A proposal for an automotive architecture framework for Volvo Cars." In 2016 Workshop on Automotive Systems/Software Architectures (WASA), IEEE, pp. 18–21.
3. Craig, B. L., & Boley, H. (2008). "Personal agents in the rule responder architecture." In *International Workshop on Rules and Rule Markup Languages for the Semantic Web*. Berlin, Heidelberg: Springer, pp. 150–165.
4. Bayer, T.J., Bennett, M., Delp, C.L., Dvorak, D., Jenkins, J.S., & Mandutianu, S. (2011). "Update-concept of operations for Integrated Model-Centric Engineering at JPL." In *2011 IEEE Aerospace Conference*, IEEE, pp. 1–15.
5. Emery, D. & Hilliard, R. (2009). "Every architecture description needs a framework: Expressing architecture frameworks using ISO/IEC 42010." *Proceedings of Joint Working IEEE/IFIP Conference on Software Architecture and European Conference Software Architecture, WICSA/ECSA 2009*. 14-17 September, Cambridge, UK, pp. 31–40.
6. Bosch, J. (2004). "Software architecture: The next step." In *European Workshop on Software Architecture*, Berlin, Heidelberg: Springer, pp. 194–199.
7. Kanjilal, D., Singh, D., Reddy, R., & Mathew, P.J. (2014). "Smart farm: extending automation to the farm level." *International Journal of Scientific & Technology Research, 3*(7).
8. Peterson, G.D., Cumming, G.S., & Carpenter, S.R. (2003). Scenario planning: a tool for conservation in an uncertain world. *Conservation biology, 17*(2), 358–366.

9. Eppinger, S.D., & Browning, T.R. (2012). *Design structure matrix methods and applications.* MIT press.
10. IDEF 0 (1993). Integration Definition for Function Modeling (IDEF0). Draft Federal Information Processing Standards Publication 183 (see also http://www.idef.com/idef0.htm).

3.9 Additional Reading

Rechtin, E. & Maier, M. W. *The Art of Systems Architecting,* CRC Press, 2010.

Buckl, S., Krell, S., & Schweda, C.M. (2010). "A formal approach to architectural descriptions—refining the ISO standard 42010." In *International Workshop on Cooperation and Interoperability, Architecture and Ontology,* Berlin, Heidelberg: Springer, pp. 77–91.

4

Modeling of Socio-Technical Systems

> The problem of communication and knowledge transfer between the various stakeholders ... primarily as a problem of the translation and structuring of system information in order to make it more digestible to all parties. This transfer of information can be greatly assisted by the use of configuration models. Such models are particularly suitable for management level investigation and understanding, as it provides a high level appreciation of a system's nature. This ensures that managers are not forced into considering a system in unnecessary depth nor requiring them to comprehend the low level operation of the components and process which are involved.
>
> **Simon Lock**
> *The Management of Socio-Technical Systems using Configuration Modelling*

4.1 Requirements of a Socio-Technical Model

An engineering system is the realization of scientific discoveries in a structured way for the benefits of human users. Examining examples such as the cochlear implant, the International Space Station, home appliances, and traffic light and road systems reveals that apart from the physical parts of a system, two other elements are required to make a system useful. These elements are part of the larger socio-technical system in which all of these systems are operating.

Figure 4.1 shows the physical element of a system as the "product" that is built using fundamental engineering sciences. This is the common view of most users, and society in general. The product is the tangible element that gives the "touch and feel" of the system. In the cochlear implant system, the implant is the product. In the traffic light system, the traffic lights, together with the controller, are the product. In software systems, the product is the program that sits in the computing device. In the commercial sense, this is what the customers feel they pay for.

Not everyone realizes that the "people" element is an integral part of the system, or that it is not limited to the users. It includes all the human participants involved to enable successful operation of the system. In the cochlear implant system, the user is obviously the patient. But who puts the device into the patient's body? Who provides regular checks and training, or system upgrades? The traffic light system also has a people element. Who decides the timing of lights and sets it up in the way the drivers see? Who provides the regular checks and repairs? In the case of the International Space Station, it is clear there are thousands of people involved in the system doing numerous tasks.

To use the product properly, a set of procedures—i.e., a process—should be defined and followed. This is the realization of practices that support proper operation of the system, "the role of engineering in systems". A defined set of procedures not only allows the people (remember there could be many people) to synchronize with the reactions of the system at different inputs during operation, but also helps the system overcome the challenges that we discussed earlier.

In Figure 4.1, the interactions of the constituents of the systems are shown by the double arrows. Without these interactions, the product is not used by people, the people do not follow the process, and the performance of the product is unpredictable without a defined process.

On top of this, the three elements are working and interacting within an "environment". If the environment is as expected by the system—e.g., the traffic light system is operating within a reasonable temperature range on a normal day—the elements in the system can work and interact correctly and perform well.

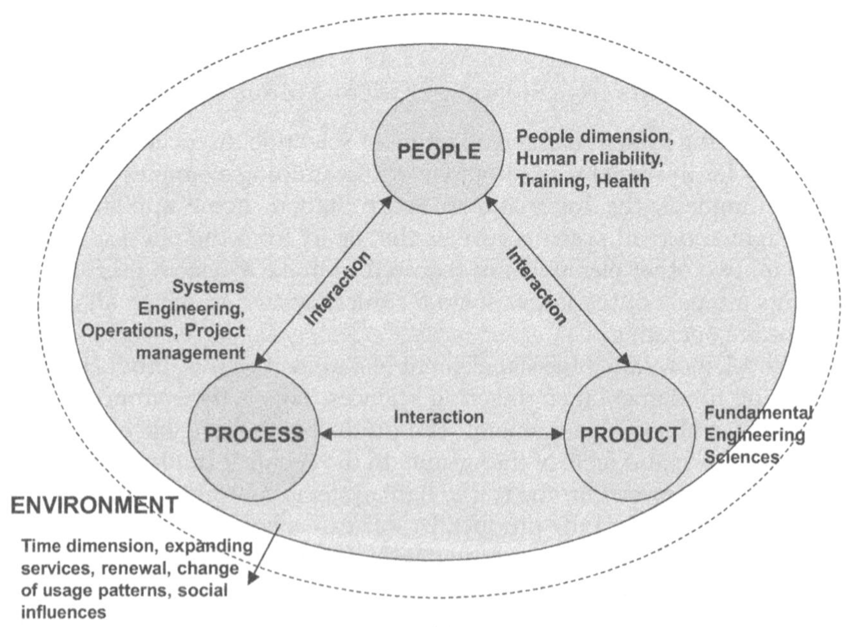

FIGURE 4.1
Elements of a system within an environment

From the perspective of the 3PE model, if the environment changes beyond the extreme conditions expected—e.g., on a very hot day the ambient temperature rises to 45° and the temperature inside the traffic lights becomes extremely high—the system will fail. To overcome this problem, the system has to change; i.e., some or all of the elements have to adapt to the new environment. If nothing is done to the system when the environment changes, it can become out of date or obsolete. All of these environmental changes are external to the business but are serious challenges for the manufacturer, because interactions with customers have to be prompt and efficient. Having identified the influence on the elements, changes can be planned carefully to ensure a holistic approach, balancing the effects on all parties while maintaining consistent business growth.

As the world's economy becomes global, many business operations face challenges to improve the quality of their products and services throughout the product lifecycle. Change in environment is irresistible. For example, particularly for major equipment with a typical lifespan of over 20 years, some manufacturers provide extended customer support that covers the product lifecycle, i.e. from the usual design oriented supplier's view to a comprehensive coverage beyond the sale and guarantee period of the product. Manufacturers have to give guidance to their customers on how to operate their products, and when and how to carry out preventive maintenance. They also must help customers identify the causes of malfunctions and how to repair equipment when it fails.

The support system should be developed for different kinds of people. A novice machine operator requires more detailed instructions than a trained one. Customers in other countries have distinct traditions and education systems, and communication may be hindered not only by language difficulties but also by cultural differences. The system should therefore be flexible and adapt to the needs of the user irrespective of any differences.

We will use an industry example to illustrate how this works.

4.2 Case Study 1

This case study involves a computer numerical control (CNC) machine manufacturer. The manufacturer has customers all over the world and hence needs to provide effective and responsive remote support to customers for the use, maintenance and troubleshooting of their equipment [1]. As the market for CNC machines has expanded, these needs have increased exponentially so that the conventional processes that were in place are no longer economically viable. In addition, the company wants to capture the vast body of expert knowledge concerning their CNC machines within and outside the company, especially as it pertains to problem diagnosis. However,

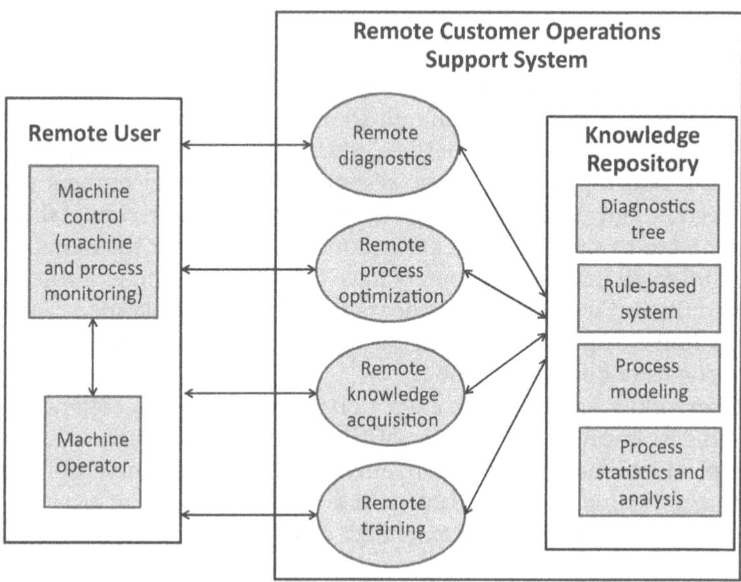

FIGURE 4.2
Functional diagram of a remote operational support system [2]

there are no ready means for gathering and storing this information, and hence no way to exploit it for remote customer support.

In this scenario, a CNC plasma cutting system manufacturer needs to create a remote customer support system which is built around a knowledge repository consisting of field and engineering information gathered from the company and their customers, as shown in the system architecture diagram in Figure 4.2. Four user interfaces—remote diagnostics, remote process optimization, remote knowledge acquisition, and remote training—are designed. These interfaces interact with the user at a remote location and interrogate the knowledge repository to provide appropriate information to the user. All functions can be easily kept up to date over the internet.

4.2.1 Remote Diagnostics

There are many non-controllable factors involved in operating a CNC plasma cutting system, e.g., changes in the conditions of the machine itself or of the input materials. The remote diagnostics system uses time-based signals available from machine sensing mechanisms to develop a remote condition monitoring system for plasma CNC cutting systems with the aim of servicing the customer anywhere in the world via the internet. The signals collected are analyzed by phase-space transformation techniques. Many fault conditions can be identified. The customer can then be alerted that something on his/her machine has high probability of failure. At the same time, the maintenance technicians are then able to make suitable preparations for maintenance to speed up repair times.

4.2.2 Remote Process Optimization

Process optimization refers to the support service which advises the user how operating parameters can be used to achieve the best results. It involves searching a knowledge base which consists of engineering design knowledge, expert knowledge in scientific investigations, past operating results and rules to decide changes to parameters.

In addition to field knowledge, there is also a need to incorporate new information that arises as the technology develops. Existing systems are generally robust in this sense because they are often designed with a particular problem in mind and hence do not need the flexibility to accommodate structural changes of the knowledge representation. To develop this capability, substantial research is then required to integrate the existing knowledge base with the new knowledge. Similarly, research is also required to investigate how changes in knowledge structure impose problems such as form of information on the delivery of the support to the customer and what solutions can be implemented.

4.2.3 Remote Knowledge Acquisition

The machines were configured as servers that communicated with the CNC machine manufacturer's global master server. Information from the operation of machines was captured from the databases of individual companies. These significantly improved sources of information enabled the product manufacturer to determine the best options for supporting the operation and maintenance of plasma cutting machines from a distance.

4.2.4 Remote Training and Online Manual

In order to keep trainees and system users engaged, multimedia authoring tools are available to create applications which can be used for self-training or as online manuals. These tools normally run on a PC and so are excellent systems for localized support. However, for customers and users who are far away and have difficulty to attend face-to-face training courses, some other form of delivery has to be employed. For example, graphics information can be successfully displayed by hypertext technology for a local support situation.

4.2.5 Modeling of Remote Operational Support Interface

The remote customer support process is in fact a loop consisting of a series of dialogues with the remote customer. These characteristics can be modeled

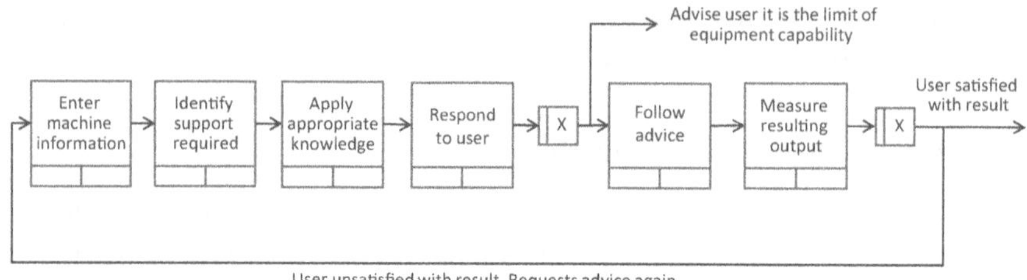

FIGURE 4.3
Remote operational support system

by IDEF3 (Figure 4.3). The system works with the knowledge base to provide answers according to feedback from the customer.

4.2.6 Mapping to the 3PE Model

In this case study, new functions were required to be developed and integrated with the product, i.e., the CNC plasma cutting machine. These functions are mapped to the 3PE model as shown in Table 4.1.

TABLE 4.1
New Functions in a Signal-Based Diagnostic System Mapped to 3PE

Function	Description	Mapped to 3PE
On-machine signal-based diagnostics capability	A new diagnostics software module based on chaotic theory and digital signal processing was developed to assist identification of faults.	Product
Communication networks and IT systems based on client–server model	The machine controller was changed from the normal standalone operating system to one that can act as a server in a network environment.	Product – People
Knowledge sharing—transform customer data into information and then into knowledge	New data-processing algorithms were developed as software modules that could process data on the machine into useful knowledge for the enhancement of operational efficiency.	People
Engineering information integration to support more effective customer service	Engineering information such as bill of material, machine configuration management, parts inventory, and resources planning were integrated from different sources including CAD, MRP, and various manufacturing sources to create a seamless operation database for the machine.	Process
The new system design requires upgrade of field products	Field upgrade for machines already installed at customers' locations was progressively rolled out according to contracted maintenance schedules.	Environment

4.3 System Capability Assessment

The 3PE model forms the basis for the assessment of the capability of a system to perform certain functionality. According to the 3PE model, a system can be viewed as three primary elements—product, process, people—and three interactions of the three primary elements, and work in a defined environment.

To develop a method of assessing the system's capability, performance indicators can be developed from the 3PE structure, as shown in Figure 4.4.

The first level of capability, CL1, is based on the 3PE model structure. The second level of capability, CL2, depends on the goal of the system. For example, a system designed to be implemented in a company for the purpose of improving processes would include focus areas labeled as A, B, C, . . . H. These process improvement focus areas are further divided into sub-areas which are derived from a review of capability factors from literature, as well as experience from similar projects implemented by the company. Rating of individual capability factors is rolled up and aggregated from individual factors in the third level of capability, CL3.

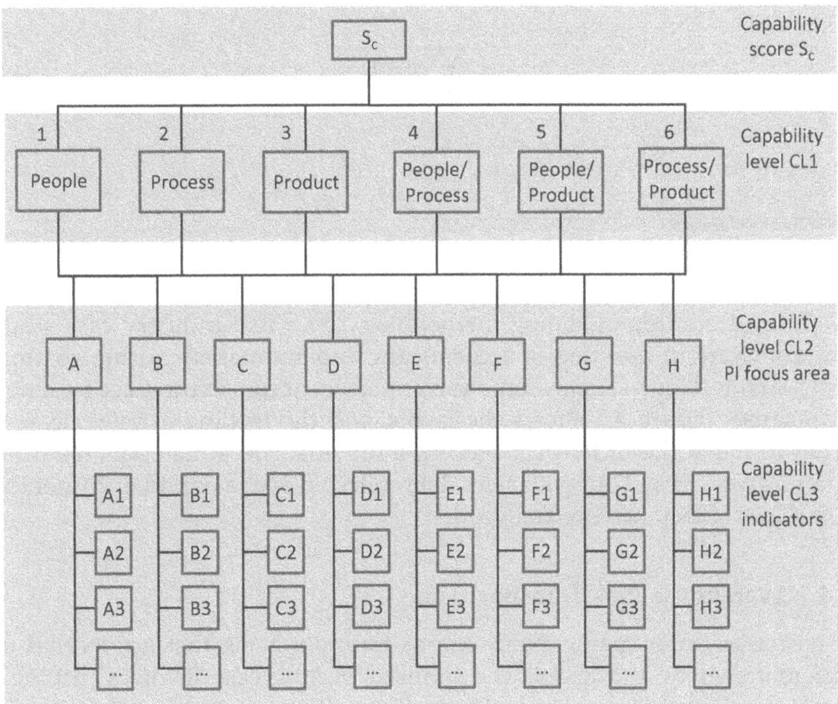

FIGURE 4.4
Capability score S_c hierarchical structure

The capability score S_c is represented as a linear weighted sum of capability indicators X_{ijk} as shown in Eqn. 4.1 where i = level CL1, j = level CL2, and k = level CL3. At each branch, the weightings sum to unity.

$$S_c = \sum_{ijk} a_i b_{ij} c_{ijk} X_{ijk} \tag{4.1}$$

Each capability factor X_{ijk} in the hierarchy obtains a rating r_{ijk} on a 1–9 Likert-type scale. This scale is 1: Non-Existent, 3: Marginal, 5: Good, 7: Very Good, 9: Excellent, and intermediate values 2, 4, 6, and 8. The ratings r_{ijk} are rescaled to a 0–100 scale for the X_{ijk} in Eqn. 4.2.

$$X_{ijk} = 100 \cdot \left(\frac{r_{ijk} - 1}{8} \right) \tag{4.2}$$

The weightings will satisfy the following equations:

$$\sum_i a_i = 1 \tag{4.3}$$

$$\sum_j b_j = 1 \tag{4.4}$$

$$\sum_k c_k = 1 \tag{4.5}$$

4.4 Case Study 2

We'll illustrate this modeling methodology by a real industry case study. The company in question is a chemical company manufacturing coatings and sealants. Due to expansion, the company acquired extra space for a new warehouse. Figure 4.5 shows the layout and the racking arrangement. In order to make the order-pick process more efficient, a materials handling analysis was carried out with sales data collected over a period of 20 months. A number of factors were analyzed.

4.4.1 Warehouse Pick Transport Time

A customer order may contain one or multiple items that are located on different shelves and aisles. The simplest picking case involves picking a quantity of items of one product type. The picker moves the trolley from its resting place near the order packing desk, picks the required quantity of an item from the racks, and returns to the order packing desk. The transport

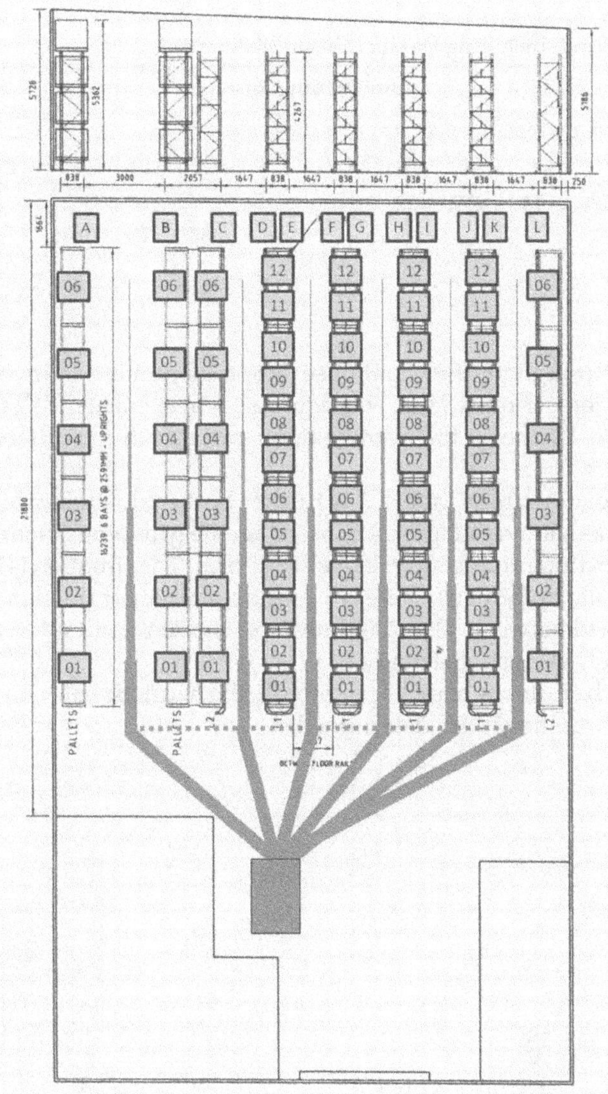

FIGURE 4.5
Aisle and bay layout in the new warehouse. Pick routes are marked from a predefined point of collection.

time required can be calculated the location of the items and the speed of travel. A more complex order involves picking multiple items types from one product group. In this case, the items may be located in only one aisle or in aisles in close proximity. The most complex order type involves the picking of multiple items from several product groups. The picks may be made

TABLE 4.2

Orders per Month Including Product Group Makeup

No. Product Groups per Order	C	P	S	T	CP	CS	CT	PS	ST	CPS	Totals
	Product Combinations										
1	199	14	215	27							455
2					0.20	14	0.15	2.6	0.05		17
3										0.65	0.65

from racks throughout the warehouse. The total number of invoiced orders recorded in the period is 9451. This equates to an average of 472.6 orders per month. Table 4.2 shows the average order per month for different product combinations.

The information from Table 4.2 is further broken down between coatings and sealant as shown in Figure 4.6. Based on the groupings made, the distribution for coating products is relatively uniform. The situation is slightly different for sealant products. There is a considerably larger fraction of products that are picked between 50 to 200 times over the 20-month interval resulting in a negatively skewed distribution.

Based on the number of picks per month, the heuristic rules for item placement in the warehouse are determined.

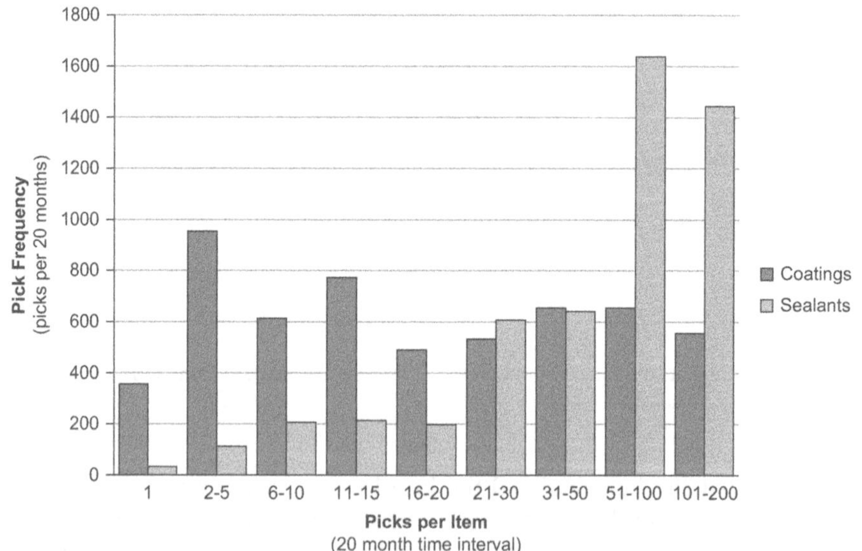

FIGURE 4.6
Total picks vs. picks per item

- First—All else being equal, place items with the highest pick rate in locations with the lowest pick transport time.
- Second—All else being equal, place items with the lowest volume per item first in locations with the lowest pick transport time.

Each rack in the warehouse has been assigned a designation Xxxy, where X = aisle designations A–L, and xxy indicates the level and location on the shelves. The product layout rules are to place products in order of priority:

1. Sealant products with 51–400 picks over 20-month interval: 36 item types.
2. Sealant products with 21–50 picks over 20-month interval: 41 item types.
3. Coating products with 21–200 picks over 20-month interval: 54 item types.
4. Coating products with 6–20 picks over 20-month interval: 175 item types.
5. Coating products with 2–5 picks over 20-month interval: 326 item types.
6. Sealant products with 2–20 picks over 20-month interval: 84 item types.
7. Coating and sealant products with one pick over 20-month interval: 391 item types.

4.4.2 Pick Time—Fixed Component (Setup Time)

Each pick has a setup time component in addition to the transport time to move the pick trolley to the rack and back to the packing desk. Setup activities are assumed as follows:

1. Walk to the pick trolley and get it operating: 1 minute
2. Search for pick item location on the data system: 0.5 minutes
3. Putting the pick trolley away after the pick: 1 minute
4. Order receipt by phone call or hand-delivery to the pick desk: 0.5 minutes

The total setup time is therefore estimated at 3 minutes per order.

4.4.3 Results of Handling Analysis

The systematic handling analysis was incorporated into a distance-traveled analysis spreadsheet as shown in Figure 4.7.

2.778	Horizontal speed (m/s)

Desk	X	Y
	10.609	-6.000

Aisle	X	Y	Distance desk to aisle (m)	Dist
A	16.232	0	A	8.223
B	16.232	0	B	8.223
C	11.852	0	C	6.127
D	11.852	0	D	6.127
E	9.367	0	E	6.127
F	9.367	0	F	6.127
G	6.882	0	G	7.064
H	6.882	0	H	7.064
I	4.397	0	I	8.637
J	4.397	0	J	8.637
K	1.912	0	K	10.566
L	1.912	0	L	10.566

Distance aisle to bay (m)

Rack type	
Pallet01	0.747
Pallet02	2.042
Pallet03	3.437
Pallet04	4.732
Pallet05	6.127
Pallet06	7.422
Pallet07	8.817
Pallet08	10.112
Pallet09	11.507
Pallet10	12.802
Pallet11	14.197
Pallet12	15.492
L101	0.747
L102	2.042
L103	3.437
L104	4.732
L105	6.127
L106	7.422
L107	8.817
L108	10.112
L109	11.507
L110	12.802
L111	14.197
L112	15.492
L201	1.395
L202	4.085
L203	6.775
L204	9.465
L205	12.155
L206	14.845

0.381	Vertical distance trolley to rack (m)
0.167	Vertical speed (m/s)

Vertical height (m)	Trolley height	Rack height	Vertical time (s)	Trolley time
Pallet4	3.886	4.267	Pallet4	23.3
Pallet3	2.591	2.972	Pallet3	15.5
Pallet2	1.295	1.676	Pallet2	7.8
Pallet1	0	0.381	Pallet1	0.0
L17	3.658	4.039	L17	21.9
L16	3.048	3.429	L16	18.3
L15	2.438	2.819	L15	14.6
L14	1.829	2.21	L14	11.0
L13	1.219	1.6	L13	7.3
L12	0.61	0.991	L12	3.7
L11	0	0.381	L11	0.0
L24	2.972	3.353	L24	17.8
L23	1.981	2.362	L23	11.9
L22	0.991	1.372	L22	5.9
L21	0	0.381	L21	0.0

Volume per slot

Rack type	Volume	Relative volume	W	D	H
Pallet	1.49	0.35	1.165	1.165	1.1
L1	0.61	0.14	1.295	0.838	0.56
L2	2.15	0.51	2.591	0.838	0.991

Assume divider is 50mm wide

FIGURE 4.7
Parameters used for transport time calculation

The systematic layout analysis improves the pick transport time from 4.9 hours per month for random allocation to 2.3 hours per month; a relative reduction of **53%**. The absolute reduction is estimated at 4.9 – 2.3 = 2.6 hours per month. This is small in the case of the warehouse in the case study due to the relatively low pick frequency. The 53% relative reduction would have a more significant effect in a warehouse with more frequent picking activity.

The estimate of 2.3 hours per month or 1.6% of the picker/packer's time may be considered negligible. Even the worst case estimate of 7.5 hours per month or 5% may be considered low. This amount of time or more can be wasted searching for product items that cannot be readily located. The most important factor to maximize labor efficiency is that the location of items be known with some predetermined certainty and the picker moves directly to the area where an item is located without reverting to a search pattern. The overall setup activity time required for each pick is more significant than the pick time, at an estimated 24 hours per month. Time savings may therefore be achieved by focusing on streamlining the setup process.

Based on the data collected, the new warehouse layout could be established. However, there are many other unknown factors. It may be more important that the location of each item to be picked is known with certainty than where the items are physically located in the warehouse. These are factors to be considered other than the product, i.e., the physical layout and item placement of the warehouse.

4.4.4 Process Improvement Capability

The capability for process improvement in manufacturing companies is dependent on many factors. In this case, the process improvement project risk is derived from the sum of eight focus areas and expanded into respective sub-areas, as shown in Figure 4.8.

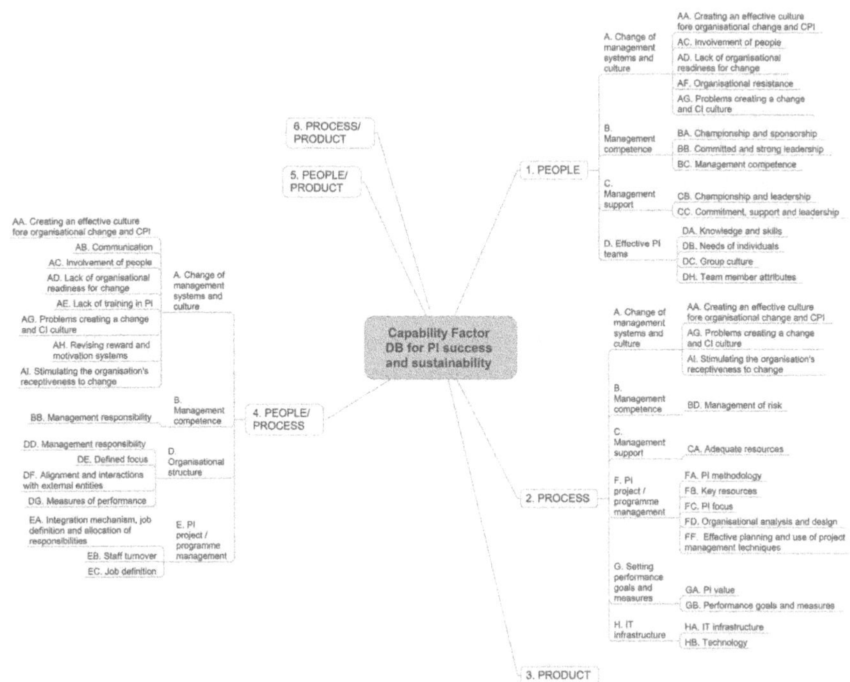

FIGURE 4.8
Hierarchy of focus areas CL2 and individual capability factors CL3

TABLE 4.3

3PE Elements and Focus Areas in the Process Improvement Project

	3PE					
	1	2	3	4	5	6
Process Improvement Focus Areas	**People**	**Process**	**Product**	**People/ Process**	**People/ Product**	**Process/ Product**
A Change of management systems and culture	X	X		X		
B Management competence in PI	X	X		X		
C Management support of PI	X	X				
D Effective PI teams	X			X		
E Organizational structure				X		
F PI project/programme management		X				
G Setting performance goals and measures		X				
H IT systems and Technology		X				

Not all elements and interactions affected a focus area in this process improvement (PI) project. The areas affected are shown in Table 4.3.

The focus areas are further broken down into sub-areas as shown in Figure 4.9, derived from the hierarchy in Figure 4.8.

The company's PI capability has been assessed with the aim of estimating how likely the company is to succeed in performance enhancement initiatives. The ratings for individual focus areas are shown in Figure 4.10.

The current estimated overall score based on information received and observed during the audit is 59. This is classified as "Very Good", as indicated in Table 4.4. A score in the "Very Good" range may be considered a minimum for achieving and sustaining results from PI initiatives. Scores in the "Excellent" range and above may be considered a minimum for achieving

TABLE 4.4

Risk Assessment Rating Scale

Score (0–100)	Rating
75–100	Excellent
63–74	Very Good
50–62	Good
37–49	Acceptable
25–36	Marginal
0–24	Unsatisfactory

	Process Improvement Focus Areas	3PE					
		1	2	3	4	5	6
		People	Process	Product	People/ Process	People/ Product	Process/ Product
A	Change of management systems and culture	X	X		X		
B	Management competence in process improvement	X	X		X		
C	Management support of process improvement	X	X				
D	Effective process improvement teams	X			X		
E	Organizational structure				X		
F	Process improvement project/program management		X				
G	Setting performance goals and measures		X				
H	Information technology systems and technology		X				

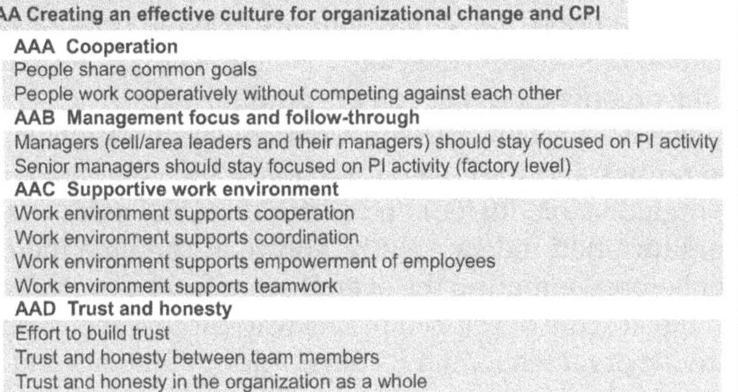

FIGURE 4.9
Elaborated 3PE focus areas

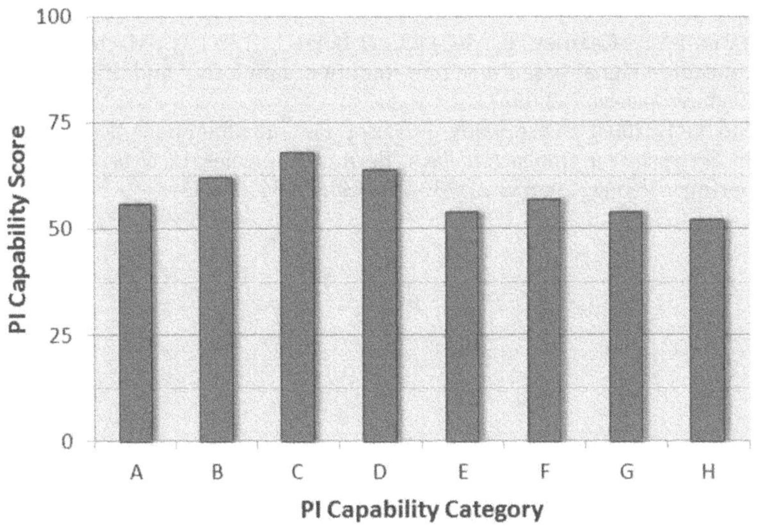

FIGURE 4.10
Rating of focus areas CL2 as aggregated from sub-areas CL3

continuous process improvement, i.e., where process improvement becomes an integral part the organization.

4.5 Reflections

The 3PE model is a starting point for the capture of information about an organization. In addition to the three basic elements of people, process, and product. In addition, the interactions are also important. Using the concepts described in this chapter, develop a 3PE model in the form of Figures 4.4 and 4.8 for the organization that you are familiar with.

Once you have established the hierarchy of the elements and interactions, use the ratings and formulae in section 4.3 to compute an overall PI score for the organization. It would be useful if several assessors could assess the same organization independently, and then combine their ratings for each element before computing the overall score for the organization.

From the PI score developed in this way, discuss the implications of such a rating to the organization for its day-to-day operations and how it is seen by customers and competitors.

4.6 References

1. Yang, S.Y.S., Kearney, T., Mo, J.P.T., & Boland, P. (2003). "Machine GP: a new concept in signal-based machine condition monitoring and diagnostics." *New Engineer Journal*, 7(2), 19–23.
2. Mo, J.P.T. (2003). "Case Study — Farley Remote Operations Support System." In *Enterprise Integration Handbook*. Bernus, P., Nemes, L., & Schmidt, G. (Eds.) Springer-Verlag, Chapter 21, 739–756. ISBN 3-540-00343-6.

Part 2

Network-Centric Operations and Matters of Context

5

Infrastructure Network Applications

> A rising tide doesn't raise people who don't have a boat. We have to build the boat for them. We have to give them the basic infrastructure to rise with the tide.
>
> **Rahul Gandhi**
> *Indian politician, circa 2014*

5.1 The Influence of Infrastructure

Many production, service, and social functions depend on the reliable and secure operation of energy supply, telecommunications, transportation, finance, and other forms of infrastructure. We have come to rely on such infrastructures to provide platforms that facilitate most everyday activities, and platform providers must integrate and support a variety of asset types in making such infrastructure available (making it their "mission"). At the same time, the evolution of new or enhanced infrastructures (such as the global positioning systems (GPS) considered in our case study) provide platforms for further innovation. The assets of interest here are generally capital-intensive and are expected to last for a long time. However, during that time, changes in user requirements, changes in technology, or changes in access to finances may result in changes in the asset and/or how it is used, which places varying demands on support services. Many of the requisite support services are knowledge-intensive, and knowledge of new technologies must be maintained in parallel with knowledge of legacy system assets.

Whilst some changes may appear to be simple in themselves, like the introduction of QR codes, they may stimulate change in the nature of interactions between financial, physical, and knowledge assets, in turn influencing who does what, and support system configurations are expected to change as a result. Interdependencies can mean that an action in one part of one infrastructure network can rapidly create global effects by rippling throughout the same network and into other networks. The potential for widespread disturbances is high. Some researchers compare this kind of behavior with that of a natural ecosystem. Others view the larger entity as a system of systems that may be both interdependent and independent to some extent. This leads to a focus on interaction and connections within both the entities themselves and within their supporting services that may span nations or the world.

In chapter 3, we introduced the idea that any functional system has four interacting generic subsystems (physical, information, decision, and knowledge), and we flow that idea on to our discussion of infrastructure. By way of example, air transport infrastructure consists of physical subsystems—aircraft and airports, information subsystems to manage bookings and plan each flight, decision subsystems operated by pilots and air traffic controllers to orchestrate operations, supplemented by the actions of people such as maintenance staff who have knowledge of each subsystem. In many parts of this book, we observe that sustainable systems must make both functional sense and business sense, and we introduce financial infrastructure as a system of systems in its own right that provides access to financial assets.

A particular kind of infrastructure may combine multiple kinds of assets to achieve its intended purpose, e.g., physical and information assets. We consider four kinds of infrastructure:

- Financial infrastructure—Bossone et al [1] describe financial infrastructure as a set of rules, institutions, and systems within which agents carry out financial transactions, and provide access to economic capital. This kind of infrastructure is represented in the operation of organizations like banks, stock exchanges, venture capital firms, and insurance companies.
- Physical infrastructure—This is the kind of infrastructure most familiar to all of us, with physical nodes like railroad hubs connected by fixed or mobile physical assets.
- Intangible infrastructure—Sharing information and codified knowledge, and reflecting rules and procedures. Some see artefacts based on codified knowledge like patents or software as an intangible asset. Edwards et al [2] refer to infrastructure software as comprising code libraries or runtime processes that support the development or operation of application software. Edwards, Jackson et al [3] describe cyber infrastructure as the set of organizational practices, technical infrastructure, and social norms that collectively provide for the smooth operation of scientific work at a distance. The internet provides the most commonly used form of intangible infrastructure.
- Intellectual infrastructure—This is about the knowledge and skills of people that may interact to learn or make decisions and get things done, all drawing on social capital (which is discussed in a later chapter). This is observed in the film-making industry, for example, where firms with technical specializations such as scriptwriting, audio management, special effects, stunts, and computer generated imagery may come together when making a particular film.

In Table 5.1, we present an overview of how these different kinds of infrastructure may draw on different kinds of assets. A development in one of these elements can transform another one. By way of example, a combination

TABLE 5.1

The Interaction of Asset Type and Infrastructure Type

Asset Viewpoint	Infrastructure Element			
	1. Financial	**2. Physical**	**3. Intangible**	**4. Intellectual**
A. Financial	Financial sector structures and combinations	Financial sector support for physical system development and operation	Financial sector support for innovation and intangible asset development	Financial sector support for knowledge-based services
B. Physical	Physical assets supporting financial sector operations	Production and services sector physical structures and combinations	Physical assets supporting the development and utilization of intangible assets	Physical assets supporting the education and entertainment sectors
C. Intangible	Intangible assets supporting the finance sector	Intangible assets facilitating the operation of physical assets	Innovation and knowledge-based structures	Intangible assets incorporating intellectual assets
D. Intellectual	Knowledge base and personnel underpinning the financial sector	Knowledge base and personnel underpinning operations and services	Knowledge base supporting the development of intangible assets	Social structures and knowledge-sharing infrastructure

of new kinds of physical and intangible assets has facilitated the evolution of digital banking (cells B1 and C1) which itself may be a platform for further developments. The shaded cells indicate some interactions observed in the GPS case study presented later in this chapter.

When we talk about a particular kind of infrastructure, e.g., transport infrastructure, we wish to make two points here. Firstly, it can be further decomposed, e.g., into related information, rail, road, sea, and air infrastructures, but each one will utilize some combination of the listed asset types. Secondly, multiple types of interactions support a particular kind of operation.

5.2 Infrastructure Development and Operation

Some researchers have noted that institutions and system procedures evolved via incremental changes in technology, markets, and regulatory processes over a long period. They warn that in the 21st century, each traditional sector is facing some degree of pressure from a number of directions:

- Discontinuity and rapid shifts in technology
- Deregulatory pressures

- Associated greater fluctuations in demand
- Natural and human threats to operations
- Unanticipated forms of competition
- Impacts of information technology on the organization and management of work
- Changing societal needs and expectations as demographics change

They see a need for research in several areas:

- Comparative analyses across technological infrastructure domains to identify common transition barriers and problems, e.g., interoperability
- Creation of integrated socio-technical infrastructure models—expanded network versions of the socio-technical systems discussed in chapter 4
- Methodology development to identify tools and approaches that support the effective transformation of infrastructure
- Application testing and evaluation

There is an interaction between two viewpoints associated with infrastructure; ideas may need new infrastructure to deploy them and new infrastructure may stimulate the development of new ideas. For example, in discussing the use of hydrogen as a clean fuel in Europe, Tzimas et al [4] noted the need for enabling large-scale distribution infrastructure, and estimated the cost involved to be about 1.5 billion euros—requiring some form of supporting financial infrastructure. Others examined the sources of infrastructure finance in Europe in terms of public procurement, project financing, and corporate sector financing. They reported a long-term trend towards more private financing was reversed over the period of the global financial crisis, and that public–private partnerships were less frequent. How might physical and financial infrastructure come together to facilitate the use of hydrogen as a clean fuel? This will depend on the attractiveness of and the momentum behind alternatives such electric vehicles.

Höyssä et al [5] studied the evolution of biotechnology development infrastructure in one region of Finland, firstly analyzing regional, national, and international interactions, and then considering the impact of establishing a physical special-purpose facility. They found this systemic view had to be complemented with a recognition of adaptations needed in the traditional relationships between some of the actors involved as the utility of social and physical infrastructure coevolved. This emphasized the point that different social actors are involved in infrastructure establishment and operation, and that their roles and the relationships between them may change over time.

Others also view infrastructure systems as large-scale socio-technical systems (see chapter 4) that do not rapidly emerge at a global scale, but rather evolve from simpler local systems through many social and technical decisions. They suggest this reflects Ashby's law of requisite variety [6], which in broad terms says that in order to respond to a diversity of problems, you need to have a repertoire of responses available. Adding something new into the mix provides possibilities for new solutions. They discuss the utilization of agent-based models to represent this situation.

Tilson et al [17] noted that ICT infrastructure has become just as important as transport, energy, and water in its influence on people's lives, work, and interaction, and explored its evolution. They described three waves, with the first continuing the earlier norm of tight coupling with industry structures, i.e., particular infrastructure for particular applications. The second wave saw the inherent flexibility of digital technologies leading to network and device convergence with some independence from traditional industry structures, e.g., through the adoption of common standards. In contrast, the third wave introduced a high level of flexibility due to the emergence of general purpose computers and adaptable software creating interdependencies with socio-technical infrastructures, e.g., the emergence of Facebook and Twitter. They note that some configurations are open (e.g., works in the public domain) and some closed (e.g., patented inventions), which influences the nature and interaction of "logical" (or code) and physical infrastructures in different social contexts. We suggest these three waves may be observed in the banking sector, which began as a collection of independent local banks, then were regulated by central banks and amalgamated (converged) or cooperated to provide regional coverage and share ATM infrastructure, and now need fewer physical facilities as people use smart cards and online facilities to organize their transactions.

Movement towards particular configurations at a particular point in time may be influenced by the interaction between various factors:

- The interconnection, overlapping, contention, and reconfiguration of physical and logical socio-technical infrastructures.
- The reconfiguration of organizational and group interconnections brought about by changes in the physical and logic infrastructures.
- Changes in the ways that new content, applications, and services are created.
- The changing locus of innovation driven by countervailing forces widening (e.g., platforms for open source collaboration) and constricting (e.g., increasing reach of intellectual property laws) the means of information creation and distribution.

In summary, there is an interaction between background social and technological changes and the evolution of infrastructure may be viewed as "waves" of change.

There is an accelerating rate of change: the growth in the use of intelligent agents emphasizes some specific aspects of infrastructure related to the interplay of intangible (e.g. communication protocols) and intellectual (knowledge) assets. The overlay of semi-autonomous smart sensors on the internet is stimulating a need for new applications, including infrastructure security, habitat monitoring, and traffic control. Technical challenges in sensor network development may include network discovery, control, and routing, collaborative signal and information processing, tasking and querying, and security. Intelligent agents draw on codified knowledge, where knowledge management infrastructure can be viewed as a knowledge-event management framework that will support structured knowledge discovery, acquisition, and maintenance.

Taking this orientation towards events further, we suggest that infrastructure development may be likened to the agile software development process, but with long iteration cycle times (years instead of days). There is a hypothesis about how a set of artefacts may be configured to meet a particular need (user story in agile software development), and this may require the development of some new artefacts (e.g., in the hydrogen fuel example cited earlier). The evolving artefacts may be assembled like a software release, which supports new or enhanced operations. Feedback from operations or the availability of new artefacts may lead to the evolution of a new "software release", upgrading infrastructure capability and/or capacity.

5.3 Infrastructure Support Systems

Zio [7] noted, "complex tasks need to be performed on modern systems in order to ensure their reliability throughout the life cycle". These tasks, characterized by Zio as maintenance and life-extension activities, "need to be properly represented, modeled, analyzed and optimized." We will expand the list of requisite support tasks and factors to be considered in their representation.

Infrastructure may be viewed as a set of components and linkages, the operation of which is to be orchestrated in response to varying requirements. This implies the existence of a performance score or script. In Table 5.2, we identify activities linking these task types with particular asset types.

TABLE 5.2

Some Characteristic Support System Scenarios

Type of Asset	Maintenance Activity	Enhancement Activity
Financial	Sustaining operations and maintenance budgets	Providing and justifying capital expenditure budgets
Physical	Maintaining physical components and links	Designing and developing physical upgrades
Intangible	Maintaining up-to-date operating procedures, maintaining software systems	Designing and developing procedural and software upgrades
Intellectual	Training to support current operational and maintenance activity needs	Learning about emerging technologies and practices

What makes sense depends on the asset lifecycle stage. For example, when analogue mobile phone networks were switched off at some point (an enhancement activity), logically similar kinds of activities were needed to support the digital network.

Over the lifecycle of an asset, different budget allocations and types of activity like training may be given different priorities. It is useful to draw on an enterprise model (see chapter 2) when identifying suitable support system architectures.

5.4 The Pursuit of Absolute Reliability

One attribute of reliable systems of systems is that no critical incidents occur, and their background operation is transparent for the provision of services. This is assured by the effective operation of a network of socio-technical support systems that continuously monitor and upgrade the physical assets, to the degree that it makes both operational and business sense.

In the pursuit of reliability by design, a review of new challenges by Zio [7] indicated that some old design problems persisted—matters of system representation and the modeling of uncertainty, but new challenges are presented by increasing system complexity, networked systems, organizational and human factors, and the reliability of software.

5.4.1 Operational Reliability Considerations

We will start with a practical example. Rochlin et al [8] observed reliability assurance practices in aircraft carrier flight-deck operations. They noted the introduction of redundancy within technical systems, and rapid availability of critical spares combined with decision/management redundancy having two aspects—firstly, internal cross-checks on decisions, even at a micro-level, and secondly, fail-safe redundancy should one management unit should fail or be put out of action. Constant multilevel communication about completed and anticipated activities was the norm, and became a familiar pattern against which anything abnormal stood out. Operating procedures represented a formal knowledge base that was supplemented by the informal knowledge base of experienced mentors working with crews that regularly changed in accordance with Navy policy. However, an enduring Navy culture has long been a source of reliability—doing the right things, the right way.

Schulman et al [9] studied the management of critical infrastructures and approaches to reliability assurance during times of crisis in electric power delivery. They also noted that the emergence of distributed systems (smart electricity grids, in their case) resulted in changed management control considerations. From observation of a number of potentially critical incidents, they noted the key role of experienced operators:

> The quest for high reliability in tightly coupled, highly interactive critical infrastructures can be characterized briefly along two dimensions: (1) the type of knowledge brought to bear on efforts to make an operation or system reliable, and (2) the focus of attention or scope of these reliability efforts. The knowledge base from which reliability is pursued can range from formal or representational knowledge, in which key activities are understood through abstract principles and deductive models based upon these principles, to experience, based on informal or tacit understanding, generally derived from trial and error.

Similar themes are observed by others in relation to the operations of other complex systems such as health management and airline operations (where briefings are held to share information, and there is redundancy—two pilots to fly the plane), and noted the value of failure simulation tools to help people learn how to respond to abnormal situations. Important support tasks identified from this discussion are designing for technical redundancy, establishing arrangements for rapid access to spare parts, asset-specific training, and context-specific creative problem-solving.

5.4.2 Reliability and Trust

Common themes in dictionary definitions of reliability are "process consistency" and "trust", e.g., from the *Oxford Dictionary* online: "The quality of being trustworthy or of performing consistently well." Spohrer and Kwan [10] make a related comment: "Without standardized measures, it is hard to agree and harder to trust."

Trust and shared values are seen as important in supporting large complex systems as implied in the previous section. A study [18] of the influence of intra-organizational trust in two high reliability organizations—a nuclear power station and an offshore oil rig—found that high levels of trust supported multilevel communications, enhancing knowledge sharing and continuous learning. Others have stressed the importance of reporting and learning from incidents, no matter how trivial they may seem, and noted the potential impact of ignoring such incidents in a high reliability context.

The reliability of operations is also influenced by inter-organizational trust, associated with supply chains, external professional services and internet–based services. However trust is a fragile management construct, taking time to build but possibly being quickly destroyed. On the basis that trust is a risky business (i.e., making assumptions about future behavior), in our consulting work we have observed that focusing discussion in potential risks (contract, competence, and goodwill risks) and how to manage them can be more productive than talking about who and what can be trusted, which brings us back to a risk-mitigation orientation. This brief discussion has identified competency maintenance, contract management, relationship management, and risk management as tasks influencing trust in a high reliability setting.

5.4.3 Reliability Assurance Tools

Simulators are used to support operator training and to explore responses to potential fault conditions in many high reliability sectors (e.g., health, aviation, power generation), but these are also complex systems requiring maintenance and updating.

Drawing on the extant literature, Hensley and Utley [11] identified reliability tools used in service operations and categorized them in relation to subsystem reliability (e.g., for failure rate analysis, control charts, fail-safe techniques, and standards), configuration (e.g., for service blueprinting and serial/parallel flows) and system reliability (e.g., failure modes and effects or root cause analyses).

In a consumer customer context, others explored the utility of combining a service blueprint model to identify potential failure points (in both front- and back-office activities), with FMEA (failure modes and effects analysis) to prioritize critical failure modes for preventative action. The reliability focus here was not so much on safety, but on service delivery. A failure was defined as any incident where customer expectations were not met. The concept was beneficially applied in a chain of hypermarket stores, where information about observed failures was provided by some 100 managers/operators, identifying seven areas for constant attention.

Suh and Ingoo Han [12] suggested an information systems risk analysis method that considers sources of discontinuity in the enterprise business model related to the specific functions serviced and the assets involved. The outcome is a risk-mitigation priority list.

The combination of drawing on accumulated experience and experimenting with future possibilities in a structured way seems appropriate in a high reliability environment, e.g., learning about the influence of extreme events from simulation studies or from opportunities to enhance reliability offered by emerging technologies and infrastructure.

The foregoing discussion has identified support tasks associated with configuration management, quality management systems, and information management.

5.4.4 Reliability in Knowledge-Based Services

The requisite support activities described so far all need knowledge and specific competencies; however, as mentioned earlier, one attribute of reliable systems is that no significant incidents are observed. So, what might be happening in the background, and how are competencies maintained? There may also be long periods between major maintenance activities, which can limit learning opportunities and make it difficult to maintain a critical mass of knowledgeable people over long asset lifecycles (e.g., 25 years or more). This can result in asset owners contracting out support to external specialists, who can maintain competency by serving a number of customers.

Potential management issues may arise in deciding to access external sources of knowledge, e.g., in creating an uncomfortable dependency. Alternatively, if an investment is made in maintaining internal expertise, this may become a source of additional business (e.g., in pursuing a servitization business model). Potential issues for buyers of knowledge-based services include the following: finding evidence supporting competence-based and contract-based trust, confidentiality matters, supplier access to supplementary knowledge networks, and measurement of the impact of the supplier's contribution. For sellers of knowledge-based services, some potential issues include communication problems arising from knowing more than the buyer, inability of the buyer to clearly specify requirements, performance perception mismatches, and seller concerns about losing outsourced skills to the buyer.

This interplay between what makes operational sense and what makes business sense when considering the pros and cons of sustaining access to requisite support knowledge highlights tasks associated with project management, boundary spanning, and knowledge management.

5.4.5 Reliability in Summary

One point to be made here is that support services continue to evolve in response to changes in the asset being supported, its operating environment, and the people involved, building on what is progressively learned. There are some recurring themes from the foregoing discussions of reliability assurance:

- There are a significant number of activities to be orchestrated.
- There is an interplay between multiple sources of knowledge and their maintenance, with interventions at multiple system levels (Table 5.3).

TABLE 5.3

Knowledge Bases Supporting Asset Reliability Management [8, 9, 11]

		Reliability Focus	
		System-Wide	**Specific Event**
Knowledge Base	Representational (formal, deductive)	Design for inbuilt reliability and redundancy, formalization of operating procedures.	Contingency scenarios, simulation exploring failure modes and effects, 'hot-spot' identification.
	Experiential (tacit; possibly trial and error)	Strategic adaptations, seeking out alternative system connections.	Reactive operations drawing on pattern-recognition skills.

- The maintenance of a strong reliability mindset provides a context for the related support systems—every element must be absolutely reliable.
- Different forms of redundancy are strategically important. This is a topic of discussion in many different professions and beyond the scope of this chapter.
- There is an interaction between support system business models and reliability assurance outcomes.

5.5 The Use of Modeling and Simulation Tools

We have referred to the use of modeling and simulation tools throughout this and other chapters. There are many books written on, and continuing research into, this subject. What we wish to present here is firstly a reinforcement of modeling and simulation in the context of infrastructure design and operation, and secondly to broadly represent the different types required.

Infrastructure facilitates the flow of money, physical objects, data, and knowledge. We have observed that in the production of goods and the delivery of services, the nature of operations and the degree of automation are shaped by flow attributes of volume and variety, and noted by Silvestro [13] in connection with services. Regardless of the application, models and simulations are used to optimize flows and identify potential bottlenecks that must be managed. As discussed in the previous section, modeling and simulation tools are also used to support operational reliability. In Table 5.4 we have provided examples of modeling and simulation activities associated with different kinds of infrastructure assets.

TABLE 5.4

A Typology of Infrastructure Modeling and Simulation Scenarios

Type of Asset	Focus of Modeling and Simulation	
	Optimization	**Reliability**
Financial	Using a Monte-Carlo method to consider factors that drive financial markets.	Exploring the risks to financial system stability.
Physical	The optimization of material flows in physical systems (e.g., a multi-terminal system for container handling).	Risk network models for evaluating project risks.
Intangible	Information flow optimization supporting decision-making	IT systems risk and vulnerability evaluation.
Intellectual	Seeking to optimize knowledge flow in particular communities.	Using training simulators to learn how to handle critical tasks and extreme events.

5.6 Case Study

GPS is an infrastructure with the mission of providing instant location information anywhere in the world, at any time of the day and under any weather conditions. In describing this case, examples of the asset viewpoint/ infrastructure type combinations will be noted in relation to the rows and columns in Table 5.1, shown by the reference [row, column].

GPS satellites were first launched in the 1960s and 1970s to support scientific research and military applications. There are three major subsystems that make up the GPS: the satellites, the ground receiving station(s), and positional data usage applications [Table 5.1 B2]. Each subsystem has evolved through the evolution of its own subsystems and other technological developments, such as the miniaturization of associated electronic components. There are interactions between these three subsystems too, e.g., when positioning accuracy is enhanced via data corrections from ground-based augmentation systems [Table 5.1 C2]. The original Russian and US systems used different communication protocols, and positioning data was commonly used in conjunction with other artefacts such as maps to support navigational decision-making [Table 5.1 C3, C4].

Civilian use of GPS infrastructure has grown explosively since the early 1990s, following initial use by surveyors and merchant ships. In 1998, Chadha [14] wrote:

> As human society enters the age of increasing mobility, location awareness becomes an important attribute of the mobile communications infrastructure. Access to location information combined with appropriate infrastructure elements will (1) enable one to find the way around and reach destinations in the most efficient manner; (2) track loved ones and provide them with emergency assistance when needed; (3) develop and get location-based services; and (4) filter the incoming information based on the location.

Some 20 years later, all of these applications are available to the broader community, with position information being incorporated into passive receivers like cameras, or into intelligent devices like smartphone navigation applications, to inform human action [Table 5.1 D2].

Initially, GPS satellites were only launched by the US and the former USSR, but other actors—China, India, and the European Union—established their own launch capabilities, albeit with some funding difficulties [Table 5.1 A2]. Satellite launching is now offered as a commercial service, and as the payload size for a particular capability decreases and the development of launch vehicles continues, costs are reducing. Cost is an important consideration because the satellites themselves have a finite life and replacements have to be launched.

Advances in ground station technology have been utilized in a number of ways. El-Mowafy [15] described a system using multiple reference stations to support aircraft precision approach and airport ground navigation. Combined with inertial system aids, high levels of accuracy and data reliability could be achieved, even if GPS signals were temporarily lost. This illustrates the principle of redundancy to enhance reliability referred to earlier in this chapter. The GPS concept is also being adapted for use in localized or indoor wireless geolocation applications [16]. Perhaps the most significant development has been the investment in low-cost receivers that are embedded in various kinds of tracking devices and smartphones [Table 5.1 A3]. Combined with other kinds of infrastructure, e.g., map data and access to the internet providing personal navigation tools, a myriad of applications have evolved.

5.7 Reflections

As mentioned in the brief discussion of the London underground rail system in chapter 1, such infrastructure is a complex system of systems. Infrastructure is seen as a form of asset, and in this chapter we have noted that the engineering and operation of complex infrastructure is supported by other kinds of complex infrastructure—financial, physical, intangible, and intellectual. This chapter emphasizes the need for trust in the reliable operation of infrastructure, which means that all four types of underpinning infrastructure must operate reliably.

Reflect on some form of infrastructure that you depend on, consider the following:

- What kinds of financial, physical, intangible, and intellectual infrastructure would be required to develop and maintain it (referring to Tables 5.1 and 5.2)?
- How might the reliability of this infrastructure be assured?
- What kinds of tools might be used in designing for reliability?

5.8 References

1. Bossone, M.B., Mahajan, M.S., & Zahir, M.F. (2003). "Financial infrastructure, group interests, and capital accumulation: Theory, evidence, and policy." *International Monetary Fund Working Paper No. 3–24.*
2. Edwards, W.K., Bellotti, V., Dey, A.K., & Newman, M.W. (2003). "The challenges of user-centered design and evaluation for infrastructure." In *Proceedings of the SIGCHI conference on Human factors in computing systems,* ACM, pp. 297–304.
3. Edwards, P.N., Jackson, S.J., Bowker, G.C., & Knobel, C.P. (2007). "Understanding infrastructure: Dynamics, tensions, and design." https://deepblue.lib.umich.edu/bitstream/handle/2027.42/49353/UnderstandingInfrastructure2007.pdf

4. Tzimas, E., Castello, P., & Peteves, S. (2007). "The evolution of size and cost of a hydrogen delivery infrastructure in Europe in the medium and long term." *International Journal of Hydrogen Energy*, 32(10–11), 1369–1380.
5. Höyssä, M., Bruun, H., & Hukkinen, J. (2004). "The co-evolution of social and physical infrastructure for biotechnology innovation in Turku, Finland." *Research Policy*, 33(5), 769–785.
6. Ashby, W.R. (1991). "Requisite variety and its implications for the control of complex systems." In *Facets of Systems Science*, Boston, MA: Springer, pp. 405–417.
7. Zio, E. (2009). "Reliability engineering: Old problems and new challenges." *Reliability Engineering & System Safety*, 94(2), 125–141.
8. Rochlin, G.I., La Porte, T.R., & Roberts, K.H. (1987). "The self-designing high-reliability organization: Aircraft carrier flight operations at sea." *Naval War College Review*, 40(4), pp.76-90.
9. Schulman, P., Roe, E., Eeten, M.V., & Bruijne, M.D. (2004). "High reliability and the management of critical infrastructures." *Journal of Contingencies and Crisis Management*, 12(1), 14–28.
10. Spohrer, J., & Kwan, S.K. (2009). Service science, management, engineering, and design (SSMED): An emerging discipline-outline & references. *International Journal of Information Systems in the Service Sector*, 1(3), 1–31.
11. Hensley, R.L., & Utley, J.S. (2011). Using reliability tools in service operations. *International Journal of Quality & Reliability Management*, *28*(5), 587–598.
12. Suh, B., & Han, I. (2003). The IS risk analysis based on a business model. *Information & Management*, *41*(2), 149–158.
13 Silvestro, R. (1999). "Positioning services along the volume-variety diagonal: the contingencies of service design, control and improvement." *International Journal of Operations & Production Management*, 19(4), 399–421.
14. Chadha, K. (1998). "The global positioning system: Challenges in bringing GPS to mainstream consumers." In *Solid-State Circuits Conference, 1998. Digest of Technical Papers*. IEEE, pp. 26–28.
15. El-Mowafy, A. (2005). "Using Multiple Reference Station GPS Networks for Aircraft Precision Approach and Airport Surface Navigation." *Journal of Global Positioning System*, 4 (1-2), 2-11.
16. Pahlavan, K., Li, X., & Makela, J.P. (2002). "Indoor geolocation science and technology." *IEEE Communications Magazine*, 40(2), 112–118.
17. Tilson, D., Lyytinen, K. & Sørensen, C. (2010). Digital Infrastructures: The Missing IS Research Agenda. *Information Systems Research*, 21(4), 748–759.
18. Babcock, H.M. (2012). A Risky Business: Generation of Nuclear Power and Deepwater Drilling for Offshore Oil and Gas. *Columbia Journal of Environmental Law*, 37(1), 63–149.

5.9 Additional Reading

Murray, A.T. & Grubesic, T. (Eds.) *Critical Infrastructure: Reliability and Vulnerability*. Springer Science & Business Media, 2007.

6

Network-Centric Operations

Network-centric warfare and all of its associated revolutions in military affairs grow out of and draw their power from the fundamental changes in American society. These changes have been dominated by the co-evolution of economics, information technology, and business processes and organizations, and they are linked by three themes:

- The shift in focus from the platform to the network
- The shift from viewing actors as independent to viewing them as part of a continuously adapting ecosystem
- The importance of making strategic choices to adapt or even survive in such changing ecosystems.

Vice Admiral Arthur K. Cebrowski, US Navy, and John J. Garstka, 1998

6.1 A Diversity of Network-Centric Operations

Network-centric operations are about making connections and sharing information. In discussing network-centric warfare, Alberts et al [1] noted that changes in military thinking had been driven by changes in the business world, moving the focus from platforms to networks and a "shift from viewing actors as independent to viewing them as part of a continuously adapting ecosystem". Wesensten et al [2] observe that in a military environment

> "networking has multiple meanings, but in the network-centric context it means computer network–based provision of an integrated picture of the battlefield, available in detail to all levels of command and control down to the individual soldier. The latter is achieved through command post, vehicle, and helmet or head-mounted displays, and individual soldier computers, all linked by radio-frequency networks."

They note that information-sharing across multiple levels of command and control is the norm, facilitating faster decision-making and self-synchronization.

Network-centric warfare is a field of study in its own right, and there are a large number of articles and books written on the subject. We will not pursue it further here, but we have suggested some additional reading for those with a particular interest in this subject.

A similar pattern of activities can be observed in civilian emergency services, where there is a higher frequency of emergent "threats", and such threats can develop rapidly. Lillrank et al [3] studied two cases involving healthcare supply chains. They suggested that process management was appropriate in situations where there was a structured flow with a sufficient volume of similar repetitions (e.g., routine GP checkups). In situations with significant exceptions, a process could be decomposed into service events that could be defined and managed as part of a flexible supply chain. A particular patient visit was described as an episode—a time sequence of events that a patient either performs or is subject to. A service event is performed by one producer (a person or a team), in interaction with a patient. It is initiated by an input that can be a request from a patient, a schedule, a signal indicating change in a medical condition, or a request from another unit. Variations are accommodated by a kind of modularity in services dynamically configured to support a particular combination of an episode and associated events. Typical events related to testing, evaluation, intervention, and monitoring activities. Considering a more structured environment, Zhang et al [4] described the use of Petri nets as tool to help optimize supply network arrangements in response to market demands for product differentiation and customization, introducing an event-oriented approach to modeling such networks.

While frequent or recurring events may be described in terms of structured flows, a network-centric view may be more appropriate if the actors involved are liable to change. By way of example, property management firms have operated tenant transaction systems for decades, supporting the electronic collection of rents, distribution of payments to vendors, and account profile maintenance for commercial and residential property leases, via a web browser and the internet. In the context of the emergent Internet of Things, Roy et al [5] considered the role of sensor networks, e.g., by combining RFID and GPS technologies, in facilitating agile supply chain management, while others have considered the matter of supply net resilience—designing a network that can respond to disruption. The term "network centric" has been coined to describe these viewpoints. With the introduction of tablet computers and smartphones, all manner of network-linked applications have evolved that facilitate day-to-day operations in a supply chain.

A well-known online retailer, Amazon.com, has grown to be one of the largest of such enterprises in the world, with separate websites for doing business in more than 15 countries. The company began by selling printed books, and subsequently developed its own e-reader product (Kindle). The company diversified into selling a wider variety of products, drawing on its established networks and network management expertise, and in 2017 acquired a large food retail chain to facilitate the online sales and rapid local delivery of food products. The company has established its own IT cloud server infrastructure and offers access to this infrastructure as a separate business. Data analytics and intelligent agents are used to personalize customer engagement events. Amazon manages highly automated picking, packing, and distribution warehouses and uses its own and other commercial logistics infrastructure to facilitate the fastest delivery pathway possible. The founder has argued that Amazon.com is not merely a retailer, but a technology company whose business model simplified online transactions for consumers.

In an e-commerce context, Romero and Molina [6] describe operational structures aimed at bringing customer networks and supplier networks together to pursue particular value co-creation opportunities. They view customers and innovators as socially responsible actors, and suggest that to compete successfully in the future companies will create value through experience-centric networks with their customers, rather than simply rely on a traditional product-centric approach.

Nambisan and Sawhney [7] viewed the emergent innovation practice of blending internal and external expertise as network centric, where a hub firm may orchestrate the utilization of such expertise. The open source software movement may be viewed in a similar way, where a multiplicity of volunteer teams work on functional elements of a large system, orchestrated by a common goal and informal protocols that have evolved over time.

The common theme is a multiplicity of connections that are liable to change in response to perceived threats or opportunities.

6.1.1 A Temporal Perspective

We see this diversity of instances of network-centric operations as similar to that observed by Silvestro et al [8] in relation to services. They developed a characterization of services based on the frequency of customer contact and contact time, and this is compared with our network-centric perspective in Table 6.1.

6.1.2 Summary

A number of themes emerge from this discussion of network-centric operations:

- Network thinking helps us maintain an overview of complex adaptive systems in a variety of situations.

TABLE 6.1

A Typology of Network-Centric Operations

Services Characterizations (after Silvestro et al [8])	Parallels in Network-Centric Operations
Low contact frequency, long contact time, high levels of **customization** with a front-office, people/process orientation are attributes of professional services such as management consultancy, field service, and bank corporate operations.	Network–centric warfare **oriented toward campaigns** where there are a series of engagements, each drawing on specialist knowledge and assets, supported by back-office supply logistics.
Moderate contact frequency, customization and discretion with a combined front-office/back-office and people/equipment orientation are attributes of a services shop such as a hotel, rental service, or bank retail operations.	Health services, exemplified as a hospital with specialist functional departments and facilities that may be **flexibly configured to meet the emergent needs** of a particular patient episode.
High contact frequency, low contact times, customization and discretion with a back-office and product orientation are attributes of mass services such as retailers and transport services.	Amazon.com, offering standard products, but with convenient access times and relatively fast delivery. Individual contact events may be extended into multiple purchase "episodes" via interaction with intelligent sales agents. Delivery relies on the **orchestration of adaptable supply and delivery networks.**

- While the focus may be on the dominant activity, e.g., a military campaign or a hospital visit, there are also two requisite support activities: developing response capabilities and anticipating future instances.
- Network-centric operations support responding to emergent events, and we suggest events be the initiation point for associated systems-of-systems design. We suggest events be represented in a hierarchical way, e.g., as events within an episode. Modeling may be facilitated by Petri net analysis (see chapter 15).
- Introducing a temporal viewpoint (Table 6.1) helps classify and better understand a particular instance of network-centric operations under consideration.
- The focus is on adaptability and responsiveness associated with fast decision-making, whatever the operating context being considered.
- If we are going to orchestrate the operation of different kinds of systems, we will need to consider potential interoperability issues.

6.2 A Focus on Decision Systems

In chapter 3, we introduced the idea that any system is made up of four interacting generic subsystems: physical, information, decision, and knowledge subsystems (see Figure 3.2). As indicated earlier, the operation of network-centric systems supports and is influenced by timely decision-making. We have indicated in chapter 3 this requires domain knowledge about the kind of decision-making needed, architectural knowledge about how the physical and information systems are working together, and about how a particular system fits into a broader ecosystem.

6.2.1 Knowledge Sharing and Decision-Making

Information and knowledge sharing between different actors is important in dealing with emergent events. Salmon et al [9] mapped the activities undertaken in a joint military-civilian emergency-response training exercise. They used an event analysis of systemic teamwork (EAST) framework to analyze the data obtained.

It was considered that the EAST approach incorporated the notion that distributed teamwork can be meaningfully described via a "network of networks" approach. EAST analyzes collaborative activities from three different but interlinked perspectives: the task, social, and knowledge networks underlying teamwork activity. Task networks represent a summary of the goals and subsequent tasks being performed within a system; social networks analyze the organization of the team and the communications taking place between the actors working in the team; and knowledge networks describe the information and knowledge (situation awareness) that the actors use and share in order to perform the teamwork activities in question. Textual analysis of team-meeting records identified many issues to be tracked in parallel, and indicated the nature of linkages between them. When they mapped tasks associated with the issues, the resultant interaction diagram looked like a plate of spaghetti. Their analysis then focused on tasks involving some uncertainty, which could inhibit operation of the network. Factors identified in this and other case studies were the organization, information management practice (potentially dealing with conflicting information), communication, situation awareness, equipment, cultural issues, a focus on tasks (compared with a desired end state) and training.

Smith and Dowell [10] made similar observations in a study of the response of multiple agencies to a UK railway incident. Their analysis focused on distributed decision-making, and described a technique they called the "progression of multiple options" to deal with emergent conditions. It was suggested that while this was effective in dealing with uncertainty, it placed considerable demands on coordination activities when time was of the essence.

Salmon et al [9] noted that most military operations use structured processes to guide decision-making during mission planning and execution. By way of example, UK military land warfare decision-making processes involve discussing seven questions in relation to a specific event:

- What is the enemy doing and why?
- What have I been told to do and why?
- What effects do I want to have on the enemy?
- Where can I best accomplish each action/effect?
- What resources do I need to accomplish each action/effect?
- When and where do the actions/effects take place in relation to one another?
- What control measures do I need to impose?

The answers to these questions are used to reach a common understanding of the battlefield situation and mission, and to choose, produce, and refine an appropriate course of action. The researchers noted a need for "cognitive artefacts" such as information overlaid on a map or diagrams on a whiteboard to facilitate coordination. Researchers in other fields have described similar kinds of artefacts as "boundary objects", where people from different disciplines describe an object or data set from their individual perspectives, leading to a much broader understanding in relation to the artefact. The general idea of identifying a set of critical questions associated with an event is analogous to the idea of asking some generic questions in respect of system and subsystem attributes presented in chapter 3.

6.2.2 Decision-Making in Different Environments

Traditional organizational decision support and strategy assumes there is an underlying order that supports rational choice [11]. A sense-making concept (described as the Cynefin framework) also considered appropriate action in unordered environments. Two ordered environments and two unordered environments were characterized as follows:

- **In a known, ordered environment,** repeatability allows for predictive models to be established and best practice identified. This is the domain of process re-engineering pursuing efficiency, where the appropriate actions are **sense-categorize-respond.** We cited an example earlier in relation to medical checkups.
- **In an ordered environment where cause and effect are knowable, but not immediately evident** (or may be known by only a few people), the appropriate actions are **sense-analyze-respond.** This is the domain of systems thinking, the learning organization, and

the adaptive enterprise, where experimentation, expert opinion, fact finding, and scenario planning may be combined. We cited an example earlier in relation to emergency hospital episodes.

- **In an unordered environment where complex relationships and interactions cause unexpected outcomes,** the appropriate actions are **probe-sense-respond.** In this environment there may be a string of cause-and-effect relationships between the agents, and both the number of agents and the number of relationships make categorization or analysis difficult. Emergent patterns may be perceived but not predicted; a phenomenon Kurtz and Snowden [11] called retrospective coherence, and see this as the domain of complexity theory. An example we cited earlier was the joint military-civilian emergency response exercise where many issues had to be tracked in parallel.
- **In an unordered environment that seems chaotic,** the appropriate actions are **act-sense-respond.** In our earlier example of a railway accident scenario, the immediate response was to pursue multiple options. This situation where multiple decision-makers observe the same phenomenon from different points of view, and "those most comfortable with stable order seek to create or enforce rules; experts seek to conduct research and accumulate data; politicians seek to increase the number and range of their contacts; and finally, the dictators, eager to take advantage of a chaotic situation, seek absolute control." Kurtz and Snowden [11] suggest that collaborating to reach consensus on a series of small actions can progress the situation.

Tay and Mui [12] drew parallels between network-centric operations in Singapore in response to an outbreak of SARS infection and network-centric warfare principles. Consistent with the Cynefin framework, they observed the need for three kinds of information to inform decision-making:

- **Abstract information** (generally tacit) drawing on experience. The question was how to map this when action-cause-effect relationships cannot be established with certainty within the short time window when it would be helpful (i.e., there is no time to see if an action is effective or not).
- **Executive information** that included trends, relationships, and deviations of groups of specific information. They suggested that this information could be analyzed to develop models and hypotheses of the situation. Assumptions and estimates could be made with varying degree of uncertainty to fill in the information gaps. Optimization with constraints, modeling, and simulation could be carried to generate possible courses of action. Tay and Mui [12] noted that while executive information may be complex in form or presentation, it could be represented and could represent the knowable.

Information in this class may be used for decision-making in complex but mostly known situations.

- **Specific information** necessary for the completion of a task or in the analysis of trends (such as number of people with symptoms in specific areas during the SARS emergency). This detailed information may be readily represented, stored, manipulated, queried, and transmitted.

They observed that decision-makers, operations coordinators, "mobile enforcers" and information content managers had to work together in different ways to support and implement the outcomes of decisions made.

In network-centric operations decision-making often involves interactions between remote contributors. Turban et al [13] considered the utility of a "collaboration 2.0" IT tool in supporting virtual decision groups. They suggested that a fit-viability combination depended on the nature of the decision task and accessible organization/infrastructure capabilities. Decision tasks identified were information sharing, idea generation and evaluation, solution selection, and solution implementation.

While having reliable information supports better decision-making, Cummings et al [14] identified ten human supervisory control challenges that could significantly impact operator performance in network-centric warfare that we see would also be relevant in other situations. These were information overload, appropriate levels of automation, adaptive automation, distributed decision-making through team coordination, complexity measures, decision biases, attention allocation, supervisory monitoring of operators, trust and reliability, and accountability. As levels of digitalization of sensor and intelligence data increase, so does the potential difficulty of dealing with large volumes of data, which may be in different formats.

6.2.3 Interacting Networks of Networks Supporting Decision-Making

In Figure 6.1, we have represented a combination of some of the ideas discussed so far. The diagram should be viewed in relation to these points:

- The decision subsystem is at the core, supported by a knowledge network and collaborating knowledge agents.
- The knowledge network supports the performance of tasks. It supports and is supported by the collection and distribution of information, and is populated by collaborating knowledge agents who may be individuals, teams, or organizational groups.
- We have used the term "social network" as an alternative to "information systems" as it supports the collection and distribution of information as well as linkages with requisite command and control structures. We have suggested that intelligence agents populate this

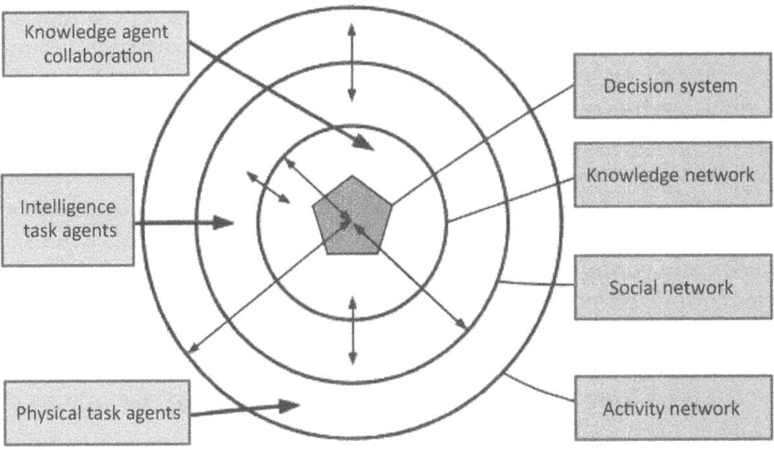

FIGURE 6.1
Networks supporting decision-making and decision-making supporting networks

network, gathering and sharing information, and these agents may be people or intelligent IT system agents.

- We have used the term "activity network" as an alternative to "physical systems", recognizing that tasks may be undertaken by physical assets, by people, or a combination of the two. In Figure 6.1, we have suggested that physical task agents populate this network, recognizing the possible variety of task agents, e.g., cyber-physical systems.

6.2.4 Summary

From a systems-of-systems perspective (physical, information, decision, and knowledge subsystems), network-centric operations bring a focus on decision-making either in responding to an emergency or in organizing alternative ways to meet customer requirements. There are a variety of decision tasks to be considered and a variety of decision environments. Fast learning is important and a particular situation may be evaluated using a structured approach to critical questioning or by experimenting with possibilities.

6.3 The Pursuit of Adaptable Connectivity

The idea of alternative network pathways has been used in telephone networks for decades. If a call cannot be connected through the most direct route, alternative pathways through multiple exchanges are used.

Such redundancy may be realized in network-centric operations, but in addition to communication between nodes, data and instructions may be sent to initiate or respond to intelligence gathering or physical activities. This requires the establishment of interoperability protocols, information security arrangements, and functionality just to manage network data flows. In dedicated systems of systems, such arrangements may be established as an aspect of total system design. Military aircraft data-bus technology, used since the 1970s, is an example of such an arrangement, where blocks of specifically configured data circulate through redundant pathways. This enables connected systems to draw on individual data elements from successive blocks or refresh them from their internal sensors. A data management system runs consistency checks on the data and manages the flow. The aircraft pilot is provided with information through displays, facilitating operational decision-making.

Network-centric operations may be organized in an analogous way. One way of facilitating the interaction of a variety of functions is through the use of standard end-to-end protocols, e.g., those that form the basis of information exchanges over the internet.

6.3.1 Use of Middleware

Another way to enable adaptable connectivity is the use of middleware, which supports the connection of disparate data sources and agents. Montenegro [15] described a space-technology middleware application allowing rerouting of connections and assignments if faults develop. Benssam et al [16] described a decision-support system providing a "digital cockpit" where the middleware includes a multi-tier IT platform providing a range of services such as data and service integration, monitoring, analysis, and process optimization. The term middleware has also been used to describe software utilities that facilitate the development of other software. Parlanti et al [17] describe the problem of integrating large volumes of data from disparate sources in relation to maritime surveillance operations and propose the use of configurable and extensible middleware offering dynamic data and systems integration capabilities.

There is a large body of literature related to the development and application of middleware in a variety of contexts, and the foregoing provides an introduction. The simple message is that when thinking about the design of systems of systems, think about the potential of middleware to support adaptable operations.

6.3.2 Dynamic Network Reconfiguration

Networks may be reconfigured to achieve the functionality required for a particular mission, making connections to available resources. Gagnes [18] introduced the topic of dynamic service discovery in a network-centric battlefield—asking what information and resources are available and where they are. It was noted that at the time of publication (2007), available commercial discovery technologies were not yet suitable, advocating for a particular intelligence gathering and discovery infrastructure and information about the current availability and viability of an asset.

Networks may be reconfigured to cope with a change in workload. Mikkilineni and Sarathy [19] discussed the issues involved in dynamically reconfiguring data center assets to be massively scalable, efficient, agile, reliable, and secure for the support of cloud-based consumer and business applications. Based on an analysis of the Intelligent Network architecture in telecommunications to identify proven concepts and key lessons, they made some suggestions:

- An evolutionary approach should be adopted.
- Providers leverage virtualization technologies.
- Providers should establish a virtual resource mediation layer that mimics the resource access arrangements developed in support of telecommunications 800 service call requirements.

Logistics networks may be reconfigured to work around blockages or to make use of a new asset.

Reconfiguration of resources rather than reconfiguration of linkages may be required. This was mentioned earlier in relation to the use of middleware to help compensate for faults that may develop in space-technology resources. Macías-Escrivá et al [20] reviewed approaches and challenges in the evolution of self-adaptive systems capable of evaluating and changing their own behavior when that evaluation indicates the intended outcomes are not being achieved, or when better functionality or performance may be achieved. An overview of a learning process model (attributed to IBM) is shown in Figure 6.2. It may be noted that embedded knowledge is at the core.

6.3.3 Summary

It may be necessary, or desirable, to dynamically reconfigure network linkages to optimize functionality provided by a particular set of resources for a particular mission, or to work around a blockage or loss of access to a particular resource. It may also be desirable to access external networks via a broader services ecosystem. Apart from managing the connections, potential interoperability issues may have to be addressed, requiring the introduction of some form of middleware. It may also be necessary or desirable to reconfigure a remote resource, and the possibilities for developing intelligent agents for that purpose may need to be considered.

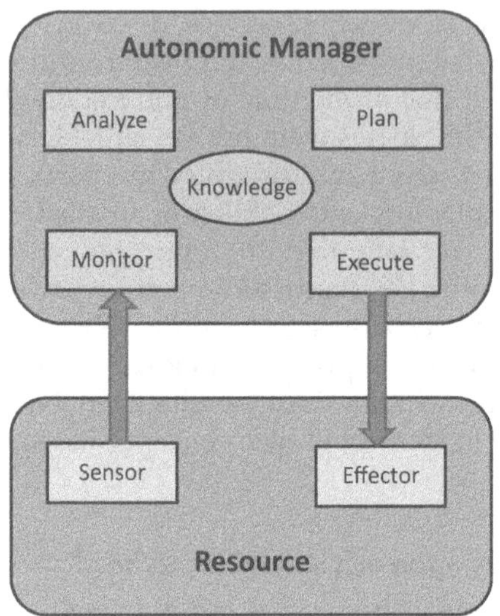

FIGURE 6.2
IBM Autonomous Computer Architecture

6.4 Sustaining Multiple Modes of Operation

While the military focus is on supporting campaigns, military organizations operate in three modes that involve the orchestration of different kinds of resource and knowledge networks:

- A **capability-development mode** where operational assets are developed and training activities are undertaken. In peacetime, this can be the dominant activity. It includes refining strategies and tactics learned from previous campaigns.
- A **capability-deployment mode** where there is relatively infrequent direct engagement with a particular threat over relatively long timeframes. This involves an iterative learning process to cope with the contingencies of a particular campaign, e.g., via the normal military practice of holding mission debriefings.
- An **intelligence-gathering and technology-development mode** aimed at being better prepared to confront potential future threats. This involves working in the imagination to learn about the implications of possible future scenarios.

While similar patterns may be observed in civilian organizations—in health emergency services operations, for example—the relative weighting given to each mode may be different, e.g., placing more emphasis on capability deployment.

Earlier in this chapter (Figures 6.1 and 6.2), we placed knowledge at the core of network-centric operations, and these three operational perspectives can be associated with different knowledge and learning domains:

- Learning from the past and internalizing such knowledge through informal interactions and practice.
- Testing the utility of past learnings and developing new knowledge through operational experience.
- Identifying new knowledge associated with external trends and breakthroughs.

Drawing on a longitudinal study by Rochlin et al [21], we suggest that an aircraft carrier at sea, where many systems of systems have to operate in harmony in different ways and at different times, provides an example of these knowledge domains operating in parallel. They reported:

> So you want to understand an aircraft carrier? Well, just imagine that it's a busy day, and you shrink San Francisco Airport to only one short runway and one ramp and gate. Make planes take off and land at the same time, at half the present time interval, rock the runway from side to side, and require that everyone who leaves in the morning returns that same day. Make sure the equipment is so close to the edge of the envelope that it's fragile. Then turn off the radar to avoid detection, impose strict controls on radios, fuel the aircraft in place with their engines running, put an enemy in the air, and scatter live bombs and rockets around. Now wet the whole thing down with salt water and oil, and man it with 20-year-olds, half of whom have never seen an airplane close-up. Oh, and by the way, try not to kill anyone.
>
> **Senior Officer, Air Division**

They noted three mechanisms that supported knowledge diffusion in the face of rapid turnover. One was the pool of chief petty officers, many of whom had long service in their specialty and had circulated around similar ships in the fleet. Another drew on time served on other carriers by many officers and some of the crew, providing some of the shared experience of the entire force. Finally, the process of continual rotation and replacement, even while on deployment, maintained a continuity that was broken only during a major refit. These mechanisms represented an uninterrupted process of on-board training and retraining that made the ship one huge, continuing school for its officers and men.

The message taken from this example is that appropriate combinations of people and technology, and informal and formal procedures can achieve remarkable results in an environment of constant learning.

6.5 Case Study

Devastating bushfires have severely impacted communities across the world, and some predict that the severity and frequency of bushfire is on the rise. Hotter weather and more intense fires requiring longer working shifts and more frequent deployments represent significant threats to the operational readiness of the personnel supporting emergency services. In south-eastern Australia, there are multiple fires every year, and very large and damaging firestorms have developed every 5 to 25 years in any particular region. After each major event there are reviews about what could have been done better to minimize the impact and reduce the loss of life. This case study draws on observations from intensive reviews of a 2009 fire in the state of Victoria and some subsequent actions taken.

Prior experience had shown that having distributed, intelligent firefighting assets deployed around the state could support a rapid response to try and contain fires before they spread. Viewing these assets as a network may facilitate combined member responses. Victorian and Californian firefighters have shared their experiences, and two Harvard University researchers reflected on command-and-control successes and failures associated with the 2009 Victorian fires. They noted that a central command-and-control facility could provide the bigger picture to cooperating emergency services and allocate resources most effectively, but could suffer from information overload and slow decision-making when confronted with fires on multiple fronts. On the other hand, a decentralized system risks having inadequate training and experience at many decentralized points of decision-making, and may allocate resources inefficiently in the absence of the bigger picture. The researchers also noted that multiple agencies operating under different jurisdictions and volunteer networks (e.g., the Red Cross) had to work cooperatively, forming temporary teams or multiple temporary network connections. It was suggested having a common understanding of information available and tasks required via a continuously evolving incident management system could facilitate such coordination (see Figure 6.1 for the underlying logic).

Victorian fire services are managed by two separate organizations: the Metropolitan Fire Brigade (MFB), comprising 47 fire stations, and the Country Fire Authority (CFA), which is supported by more than 1000 local volunteer brigades. The operations are distributed to provide a relatively quick response to the identification of a fire. Both organizations maintain central control centers, and can support each other when needed. Coordination with police, ambulance services, and state emergency services group that can undertake search and rescue missions is also maintained. The individual units maintain (and in the case of the CFA, are also supported by) close community links. For the MFB, action can be initiated by automated building alarms or by telephone calls to an emergency number.

It is not uncommon for neighbors to be the first to spot a residential fire that may start in a roof or a garage. This may also happen in the CFA regions, but here, fires starting in remote forest areas, e.g., by lightning strikes, can be difficult to detect and access. At the height of the bushfire season there can be 50 fires burning simultaneously in the state, which can challenge coordination of resources.

The Government of Victoria has established a public emergency management website with tabs having the headings "Being prepared", "Incidents and warnings", and "Relief and recovery", with pull-down menus under each type of event pointing to fire, flood, storm, earthquake, tsunami, thunderstorm asthma, and extreme heat events. Current incidents are displayed both as a list and as icons on a map. This presents an example of the cognitive artefacts that synchronize the understanding of different stakeholders [9].

The three headings may be viewed as different modes of operation. Under the heading of preparedness there is advice for citizens about being prepared for residential fires, e.g., checking smoke detector alarms, and details of actions being taken by fire authorities to minimize risk. These include training drills and controlled burns in targeted potential bushfire areas to reduce the amount of fuel in the forest understory. This practice has emerged from prior experience that while a fire may spread rapidly through the treetops, it is most likely to be sustained by burning on the ground. Under the heading of incidents and warnings there is current information about the location and size of fires, and advice for citizens, e.g., to head to evacuation centers. Firefighters may adopt a variety of strategies, e.g., calling for additional resources, including aerial water or retardant bombing, or "back burning"—deliberately starting a fire ahead of the major front that will be drawn into that front by the severe updrafts associated with it, slowing its progress. Under the relief and recovery tab there is advice about when it is safe to return, the status of particular areas, and details of support available through a variety of agencies.

Both Victorian firefighting organizations have been long-term participants in a national cooperative research center charged with accumulating knowledge about fires and other kinds of incidents, learning from the past, exploring improvements in response practices, and identifying future options. Technology tools include satellites, aircraft and drones equipped with optical and infrared sensors, and land and aerial firefighting vehicles. As more sources of intelligence are added, data analytics has become an important supporting technology. Several hundred studies have been completed and reported, including studies of management strategies, human factors, and technology tools such as remote sensing and "big data" analytics.

Human factors include the need to share the knowledge of experienced firefighters and community members, and the need to encourage an ongoing flow of volunteers as rural demographics change. Understanding social networks and their design is an essential aspect of maintaining the current capability, where the local CFA station can act as a community hub, and community donations can fund the acquisition of equipment. Social media data mining can rapidly assemble information from multiple community observations about a fire.

All four decision-making environments discussed in section 6.2.2 may be observed in the different events to be serviced:

- There may be a small fire that can be dealt with following practiced procedures.
- A fire in a remote region may be identified, but the appropriate response may depend on learning more about it and considering options available.
- In the coordination of responses to multiple fires that stretch resources and information about the situations change rapidly, there may be a string of cause-and-effect relationships between multiple agents.
- The emergence of large firestorms can introduce an apparently chaotic situation where the immediate focus is on action to save human life while testing the effectiveness of multiple response activities.

6.6 Reflections

Network-centric operations are supported by extensive information-sharing associated with "service events". There is a typology of operations that represent different combinations of volume and variety in relation to such events. From your personal experience, can you identify a network-centric operation that responds to infrequent, complex events and one that responds to frequent, relatively simple events? What is needed to support decision-making in each example? How might network connections be required to adapt in responding to each type of event? Can you identify different modes of operation for preparing for the event, enacting the event, and anticipating future events?

6.7 References

1. Alberts, D.S., Garstka, J.J., & Stein, F.P. (2000). *Network Centric Warfare: Developing and Leveraging Information Superiority.* Assistant Secretary of Defense (C3I/Command Control Research Program) Washington DC. http://www.au.af.mil/au/awc/awcgate/ccrp/ncw.pdf (last accessed August 2018).

2. Wesensten, N.J., Belenky, G., & Balkin, T.J. (2005). "Cognitive readiness in network-centric operations." Walter Reed Army Inst of Research, Silver Spring MD Div of Psychiatry and Neuroscience. Available in *Parameters: Journal of the US Army War College*, 35(1):13.
3. Lillrank, P., Groop, J., & Venesmaa, J. (2011). "Processes, episodes and events in health service supply chains." *Supply Chain Management: An International Journal*, 16(3), 194–201.
4. Zhang, X., Lu, Q., & Wu, T. (2011). "Petri-net based applications for supply chain management: an overview." *International Journal of Production Research*, 49(13), 3939–3961.
5. Roy, S., Jandhyala, V., Smith, J.R., Wetherall, D.J., Otis, B.P., Chakraborty, R., & Sample, A.P. (2010). "RFID: From supply chains to sensor nets." In *Proceedings of the IEEE*, 98(9), 1583–1592.
6. Romero, D. & Molina, A. (2011). "Collaborative networked organisations and customer communities: value co-creation and co-innovation in the networking era." *Production Planning & Control*, 22(5-6), 447–472.
7. Nambisan, S. & Sawhney, M. (2011). "Orchestration processes in network-centric innovation: Evidence from the field." *The Academy of Management Perspectives*, 25(3), 40–57.
8. Silvestro, R. (1999). "Positioning services along the volume-variety diagonal: the contingencies of service design, control and improvement." *International Journal of Operations & Production Management*, 19(4), 399–421.
9. Salmon, P., Stanton, N., Jenkins, D., & Walker, G. (2011). "Coordination during multi-agency emergency response: issues and solutions." *Disaster Prevention and Management: An International Journal*, 20(2), 140–158.
10. Smith, W., & Dowell, J. (2000). "A case study of co-ordinative decision-making in disaster management." *Ergonomics*, 43(8), 1153–1166.
11. Kurtz, C.F. & Snowden, D.J. (2003). "The new dynamics of strategy: Sense-making in a complex and complicated world." *IBM Systems Journal*, 42(3), 462.
12. Tay, C.B. & Mui, W. K. (2004). "An architecture for network centric operations in unconventional crisis: lessons learnt from Singapore's SARS experience." Thesis of Naval Postgraduate School, Monterey, CA, USA.
13. Turban, E., Liang, T.P., & Wu, S.P. (2011). "A framework for adopting collaboration 2.0 tools for virtual group decision making." *Group Decision and Negotiation*, 20(2), 137–154.
14. Cummings, M.L., Bruni, S., & Mitchell, P.J. (2010). "Human supervisory control challenges in network-centric operations." *Reviews of Human Factors and Ergonomics*, 6(1), 34–78.
15. Montenegro, S. (2009). "Network centric core avionics." In *Advances in Satellite and Space Communications, 2009 (SPACOMM 2009)*, IEEE, pp. 197–201.
16. Benssam, A., Berger, J., Boukhtouta, A., Debbabi, M., Ray, S., & Sahi, A. (2007). "What middleware for network centric operations?" *Knowledge-Based Systems*, 20(3), 255–265.
17. Parlanti, D., Paganelli, F., & Giuli, D. (2011). "A service-oriented approach for network-centric data integration and its application to maritime surveillance." *IEEE Systems Journal*, 5(2), 164–175.
18. Gagnes, T. (2007). "Assessing dynamic service discovery in the network centric battlefield." In *Military Communications Conference, 2007 (MILCOM 2007)*, IEEE, pp. 1–7.

19. Mikkilineni, R. & Sarathy, V. (2009). "Cloud Computing and the Lessons from the Past." In *Enabling Technologies: Infrastructures for Collaborative Enterprises, 2009, WETICE '09, 18th IEEE International Workshops on*, IEEE, pp. 57–62.
20. Macías-Escrivá, F.D., Haber, R., Del Toro, R., & Hernandez, V. (2013). "Self-adaptive systems: A survey of current approaches, research challenges and applications." *Expert Systems with Applications*, 40(18), 7267–7279.
21. Rochlin, G.I., La Porte, T.R., & Roberts, K.H. (1998). "The self-designing high-reliability organization: Aircraft carrier flight operations at sea." *Naval War College Review*, 51, 97–113.

6.8 Additional Reading

Cares, J. (2006). *Distributed networked operations: The foundations of network centric warfare*, iUniverse.

7

Social Networks and Creative Collaboration

> If you have an apple and I have an apple and we exchange these apples then you and I will still each have one apple. But if you have an idea and I have an idea and we exchange these ideas, then each of us will have two ideas.
>
> **George Bernard Shaw, 1856–1950**

7.1 Social Networks and Social Capital

Social networking has been cited as an activity that facilitates innovation and growth in SMEs, access by small businesses to international markets, and community action to confront environmental issues. IT tools may facilitate such activities, and many firms, both large and small, have established Facebook accounts as a means of interacting with their customers and utilize social networks to facilitate open innovation.

Social networks are regarded as an intangible asset, and a significant element in the building of social capital. Social capital has been described as providing the glue (ties) and the lubricant (providing access to resources) that facilitates cooperation, exchange, and innovation. To possess social capital, a person must establish relationships with others, and it is those others, not himself or herself, who are the actual source of his or her advantage. There is continuing debate about the characteristics of social capital, but the following set is derived from a number of academic sources:

- Relational capital—seen as having three component levels. The first is described as bonding capital—links to people based on a sense of common identity ("people like us") such as family, close friends, and people who share our culture or ethnicity. Communities of practice are seen in this category. It requires mutual understanding and is typically associated with strong and enduring ties. The second is described as bridging capital—links that stretch beyond a shared sense of identity, e.g., to distant friends, colleagues, and associates. It requires an ability to accommodate difference and is typically associated with weak and intermittent ties. The third may be described as hierarchical capital—links to people or groups further up or lower down the social ladder, and is reflected in the commonly used expression "it's not what you know but who you know".

- The expectation of reciprocity—where donors provide privileged access to resources in the expectation that they will be repaid in some way at an unspecified time in the future.
- The concept of norms and sanctions that condition behavior.
- Matters of reputation, including the establishment and maintenance of trust, or the establishment of a trusted "brand".

Management researchers have found that the presence of social ties supporting links with external team members positively influenced commercial outcomes realized by inventive teams. They explored the underlying mechanisms of the types of collaboration adopted, and found evidence of benefits from both knowledge diversity and from having team members from multiple institutions, potentially accessing an extended network.

7.2 Why Collaborate?

There is more than one dictionary definition of collaboration:

- "The action of working with someone to produce something, or to cooperate with an agency or instrumentality with which one is not immediately connected." Some synonyms for the term relate to a process of collaborating that requires particular skills (e.g., teamwork), and some to collaboration as an entity (e.g., an alliance).
- "Traitorous cooperation with an enemy." Synonyms for this definition include collusion, which introduces the consideration of matters of ethics and trust as an aspect of collaboration. Some legal frameworks outlaw collusion in business as a means of gaining unfair competitive advantage.

Themes common to both definitions are the intention to work towards a common goal, sharing knowledge, and working efficiently.

Growing a business or obtaining a competitive advantage is an economic goal. A social goal might be to deal with a local or global problem, e.g., related to health management. Protecting a particular ecosystem could be an environmental goal. To achieve these goals, collaboration may be necessary to build scale or to undertake a broad range of activities.

Many governments are promoting knowledge and resource sharing between industry, academics, and government organizations as a means of achieving such goals, and support intermediary organizations that facilitate suitable collaborations. Interventions may include incentives and rewards to stimulate collaboration.

Getting things done efficiently may involve focusing on what you do well and outsourcing other activities. Establishing collaborative arrangements with outsource partners can bring mutual benefits compared with simple arms-length contracting. How does this relate to the engineering and operation of systems of systems? In this context, the development of interaction linkages, protocols, and interoperability requirements facilitate efficient operations, and potentially provide a foundation for the establishment of intelligent agents operating in the Internet of Things described in chapter 8 in this book. In the next section, we introduce some attributes of a system operating environment.

7.3 The Contingent Nature of Collaboration

As noted earlier, some collaborations aim to build scale, and in this circumstance, the firms involved may have similar capabilities, but may be traditional competitors. There may be some apprehension about working together, but the firms are likely to have common understandings of industry norms and customer requirements. They may only be prepared to share a limited amount of information. It may take some time to work out how the firms may complement each other, but when this is achieved, effective operations follow quickly.

Where the aim is to expand the scope of tasks undertaken by bringing firms with complementary capabilities together, there may be no such inhibitions, but the collaborators may not have common understandings of industry norms and customer requirements, and they may not appreciate the relative level of competency of each partner. Addressing these matters may lead to progress stalling after a flying start. These firms need to find ways in which they are similar.

As in the development of any kind of system, both kinds of venture require a persistent champion who can see the ultimate benefit of the collaboration and may need the services of an intermediary knowledgeable about the industry requirements to be met and about the operation of collaborative ventures. While every collaboration may be unique in some way, firms with prior collaboration experience are likely to be better placed to participate and extract value from a collaborative venture.

Regardless of the collaboration goal, individual participants may have different engagement attitudes and objectives, as illustrated in Table 7.1.

Collaborations may have an exploratory orientation, developing ideas as options in a pre-competitive stage, or may have an exploitation orientation,

TABLE 7.1

Characteristics of Different Types of Engagement

Nature of Engagement	Common Goals	Inter-Dependency	Nature of Trust	Knowledge Sharing	Procedures
Collaboration	Joint goals, joint responsibilities, creating together	High within agreed scope	Relationship based, evident reciprocity	Open within the scope of the collaboration	Mutually agreed
Cooperation	Compatible goals with individual entities working apart	Case-specific	Case-specific, but contract and competency-based as a minimum	Requisite only within scope of cooperation	Mutually accepted
Coordinated networking	Complementary, aligning activities for mutual benefit	Limited, some dependency on the intermediary	Trust in intermediary	Conditional exchange	Mutually understood
Networking	Communication & information exchange	Low	Minimal quantum assigned	Broadcast general information	Informal codes of behavior
Arms-length contracting	Transaction-based	Dependent on relative power	Contract & competency-based	Requisite information only	According to contract

combining resources to deploy and/or expand access to a new market. Quite different skill sets are required in each case.

Some collaborations are stimulated by government interventions aiming to build regional capability or addressing a social issue. This brings with it sets of rules and reporting arrangements to be satisfied, and many of the targeted smaller firms are not well placed for this, creating a role for a different kind of intermediary.

7.4 Can Collaboration Be Designed?

The view taken here is that multi-partner collaborations in particular can be viewed as a network form of organization, and there have been studies of different collaborations from multiple viewpoints. In the following parts of this chapter, we consider collaborative ventures in the context of an ISO 42010 framework (see chapter 3), building on the foregoing parts.

At a generic level, the mission of a collaborative venture may be broadly represented in terms of one of four scenarios shown in Table 7.2. Even at this

TABLE 7.2

Mission/Goal Orientations of Different Collaboration Scenarios

	Exploration Orientation	Exploitation Orientation
Building scale	Example: Multiple distributed research centers acting as a larger virtual enterprise under a common theme such as understanding nanotechnology	Example: Multiple mining sector SMEs marketing products and services globally under the umbrella of a common brand and sustaining a reputation for excellence
Expanding scope	Example: The search for social problem solutions requiring contributions from multiple medical science disciplines combined with multiple social science disciplines	Example: Strategic alliance between large pharmaceutical enterprise and new technology start-up company combining new inventions with large-scale complementary assets

coarse level of representation, clear differences between the scenarios can be recognized.

A number of functional systems have to be established to facilitate collaboration, each having its own protocols. Three kinds of collaboration leadership are needed: one enthusiastically and persistently promoting the collaboration, one supporting collaboration as a strategic tool and negotiating for resources, and a third performing a broker function, negotiating with a disparate group of stakeholders.

At an operational level there are also three roles to be enacted, each of which can be associated with an aspect of social capital:

- The first functional role required is the communicator. Individual enterprises have their own protocols for reporting on operational events and issues, and on influential changes in the external environment. In our experience, a collaborative venture has to work harder at communications, simply because more stakeholders at different levels from different professional or corporate cultures are involved. As the stakeholders are most likely remote from each other, supporting ICT tools will be required. This function draws on hierarchical linking social capital.
- The second functional role is the coordinator—managing tasks and building teams, and developing bonding social capital. Generic tasks to be organized may be classified as projects or events. An issue to be addressed here is the kinds of project management tools that will be used and the kinds of tools that will be used for technical work, as each participant may use their own kinds of tools at their home base.
- The third functional role is the integrator, managing relationships—a kind of career diplomat role building bridging social capital to deal with issues arising from the different capabilities and objectives of the collaboration stakeholders.

A wide variety of organizational architectures have been observed in collaborative ventures. One form can be described as an *ad hoc* collaboration where protocols for working together have been established in advance, but an operation is only assembled as needed to organize an event or confront an issue. The term "tiger team" is sometimes used within larger organizations to characterize such a short-term collaboration. Longer term dyadic collaborations have been described as strategic alliances, but such alliances may also involve more than two participants.

Other forms of organization can be characterized according to their aim:

- A goal-oriented network established to grasp a specific opportunity, or to support ongoing operations in some way. An example of the latter is the Amul milk company, which is organized around a hierarchy of regional and local cooperatives in India and is jointly owned by more than 3 million milk producers of various sizes.
- A long-term strategic network, established to facilitate ongoing professional interaction, or a flow of opportunities. This latter form of organization has been given the name "breeding network" by some researchers, with industry clusters and collaborative virtual laboratories being cited as examples.

Another aspect of architecture is the configuration of the actors. Some social network analyses show a galaxy formation, with linkages between multiple hub actors (the brighter stars) and less influential actors. Other analyses reveal a sun/planet configuration with smaller planetary actors circulating under the influence of a central dominant actor.

We regard all forms of collaboration as projects with their own lifecycle. Even long-term strategic networks will either disband or be reconfigured as external conditions and the circumstances of their stakeholders change. It is our experience that specifying the combination of mission, functional system requirements, and generic architecture at this level of abstraction provides the foundation for designing a collaborative venture, recognizing it will evolve through stages. This logic has similarities with some of the enterprise architecture models described in chapter 2.

7.5 Generic Stakeholders and the Identification of Concerns

Collaborations are established between government actors (e.g., local and regional authorities), between academic actors (e.g., universities in different countries), between commercial actors (e.g., large and small firms), or between community actors (e.g., different lobby groups). Or, collaborations may bring some or all of these kinds of actors together (e.g., for the common good of a region, or to unite against a common "enemy"). Actors that would normally be seen as competitors may work together in such circumstances.

Just as we drew on the social sciences to introduce the idea of social capital, we draw on another concept from the social sciences called structuration theory to discuss some generic concerns.

Social scientists have observed two-way interactions between social agents and social structures. They suggest that the actions of agents are influenced by the social structures they are embedded in. But at the same time, working within these structures validates their utility, and agents may choose to change structures that do not meet their requirements. (This is called the duality of structure where structure is both a medium and an outcome of social activity.) The proponent of this theory, Giddens [1], characterized three aspects of social structure:

- Structures of signification—Facilitating communication and shared meanings—how do we understand the significance of actions, transactions, and relationships.
- Structures of legitimation—Norms and rules that facilitate working together—what is sanctioned and what is not.
- Structures of domination—Exercising power and allocating resources—showing leadership while exercising control. Some matters of governance are discussed in chapter 9.

We will discuss each in turn.

7.5.1 Structures of Signification

We mentioned the need for a communication function and shared understandings between all internal and external stakeholders. If this is not effectively implemented, some stakeholders may not appreciate the significance of the collaboration.

7.5.2 Structures of Legitimation

Earlier in this chapter, we identified the establishment of norms, rules, and expectations—aspects of organizational or regional culture—as an element of social capital. Concerns may emerge if there are significant differences in organizational culture to be accommodated, putting pressure on the integration function. Expectations of reciprocity and trust were also mentioned as elements of social capital. If these behaviors are not evident, a collaborative venture may fail unless the integration function can adequately promote them. In studying the operation of B2B networks, some European

TABLE 7.3

Some Particular Concerns Influencing Participation

	Purpose	Engagement	Similarity	Maturity	Outcomes
Network pictures	Coherence and utility of activities	Antecedents	Predisposition & prior experience	Predisposition & prior experience	Relative importance
Networking activity	Focus of activity	Scale and scope of collaboration	Scale and scope of collaboration	Scale and scope of collaboration	Delivery capability
Network outcomes	Relative importance	Learning capability	Delivery capability	Learning capacity	Resource adaptability

researchers observed that the legitimization of participation in a collaborative venture was influenced by the perceptions of key managers in the stakeholder organizations. They observed that managers were interested in the nature of the ecosystem beyond a proposed collaboration and their position in it—described as "network pictures" by the researchers. They were interested in the nature of the activities their firm might be expected to undertake, and the outcomes from participation. The researchers identified some generic concerns to be considered from each viewpoint, illustrated in Table 7.3.

Another aspect of legitimation—more "rules"—relate to trust. Working with partners that do not trust each other or building trust in the collaborative entity can be problematic. We have observed corporate buyers not wanting to do business with a virtual entity comprising members organized in the galaxy formation mentioned earlier. They wanted to contract with an established company that took responsibility for organizing others—in effect supporting the sun/planet configuration. Trust is a fragile construct. It takes time to build, but it may be rapidly destroyed. These are not the attributes of a robust management practice, and in seeking to rapidly establish a new collaboration, there may be little time to build trust. A level of trust may be initially assigned at the personal level; for example, from face-to-face meetings of CEOs, who may rely on information about individual experience and reputation. At the personal level there is an assumption that those trusted are free to make their own decisions. However this may not be valid at the enterprise level, where organization "rules" may intrude. Trust is based on perceived similarities with past situations, and on some personal feedback process indicating positive or negative outcomes. Alternatively, this feedback process may initially be based on the experience of others, and such information takes time to acquire. Both of these bases require some history to exist. An example is the star rating given to a hotel by previous guests, where the credibility of the rating is also related to the number of assessors. Drawing on studies of B2B relationships in global automotive supplier networks, Sako [2] proposed a three-component model of trust: contract-based trust—will I do what I say I will do; competence-based trust—can I do what I say I will do; and goodwill-based trust—will we each act in the other's interest. Sako also suggested that certain organizational competencies are needed for the development of trust.

7.5.3 Structures of Domination

A Harvard Business School briefing paper [8] describes sources of power arising from position (able to legitimize an initiative and allocate resources), from relationships (know-who), and from personal capabilities (know-how).

Booher and Innes [3] note that collaborative policy planning can result in network power to support its implementation. They suggest that three conditions govern the relationship of agents in a collaborative network: diversity, interdependence, and authentic dialogue. These networks are seen to behave like complex adaptive systems where the network as a whole is more capable of learning and adaptation in the face of fragmentation and rapid change than a set of disconnected agents. Githens [4] observed that notions of collaborating and exercising power are founded in different philosophical perspectives, and that respecting the need for autonomy while understanding the need for community may require leaders to both stimulate critical questioning and exercise power in reconciling differences.

Failure to recognize the relative merits of "power over" vs. "power with/assigned, e.g., in allocating resources, may influence the dynamics of a collaboration. Those who feel powerless in a collaboration may limit their participation, while those who feel they are in a dominant position may not feel a need to collaborate. In a study of social networking in nursing, Raatikainen [5] found that nurses who had enough power and knowledge to facilitate change possessed both wider and deeper cognitive and moral dimensions and had better skills in human interaction than did the "powerless" nurses. In a study of supply chain collaborations, Sridharan and Simatupang [6] suggested that power and trust play a critical role in creating a mutual competitive advantage of collaboration, noting that some participants may exercise power in an opportunistic way to appropriate value for themselves rather than creating better value. We have also observed this in some of our case study work.

We indicated earlier that powerful customers can influence the way a collaboration is configured, insisting on only working through one collaboration member.

Taking an entirely different perspective, it has been observed that in a social justice system, building collaborations can be an effective means of dismantling unwanted structures of domination (e.g., associated with criminal gangs).

7.6 Viewpoints and Models

We suggested earlier that collaborations might be viewed as discrete projects that have generic lifecycle stages. Collaborations also have distinct lifecycle stages:

- Creation, involving initiation and establishment of the foundational arrangement.
- Operation, with parallel evolution building on what is learned from operations, and adapting to changes in the internal or external environment.
- Dissolution, when a collaboration goal is achieved, or metamorphosis, where those involved may pursue a new purpose together.

Different actions and capabilities are needed at each stage.

There are working arrangements to be considered within the collaboration network, in relation to linkages between the network and its operating environment. In the following we represent these internal and external viewpoints and aspects of each, drawing on extracts from extensive research into collaborative network organizations (CNO) completed in Europe in 2007 (see additional reading).

Starting with the internal viewpoint, four generic elements may be represented to varying extents as four subcategories of the viewpoint:

- Active entity—a tangible object that can perform an action in the CNO, e.g., a member/partner organization or individual.
- Passive entity—a tangible object that cannot behave and/or perform any action in the CNO; rather it is a resource that supports actions, e.g., an information or ICT resource.
- Action—a procedure or operation that is executed within the CNO, e.g., the formal registration of a CNO member, competency management, contract negotiation, and conflict resolution processes.
- Concept—an intangible aspect in the CNO that can be also associated with active/passive entities or actions reflected in practices and policies, e.g., an assigned role (associated with an organization in the CNO).

The four subcategories are:

- A Structural Dimension. This dimension addresses the structure or composition of the CNO's constituting elements (namely its participants and their relationships) as well as the roles performed by those elements and other compositional characteristics of the network nodes such as the location, time, etc. This perspective is used in many disciplines (e.g., systems engineering, software engineering, economy, politics, cognitive sciences, and manufacturing), although with different "wording" and diversified tools.

- A Componential Dimension. This dimension focuses on the individual tangible/intangible elements in the CNO's network, e.g., the resource composition such as human elements, software and hardware resources, information, and knowledge. Not all these elements are "physical" in a strict sense; in fact some, e.g., knowledge, are conceptual, but rather together they represent the "things" out of which the network is built. Furthermore, the componential dimension also consists of the ontology and the description of the information/knowledge repositories that pertain to the CNO.
- A Functional Dimension. This dimension addresses the base functions and operations available in the network, and time-sequenced flows of executable operations (processes and procedures) related to the different phases of the CNO lifecycle.
- A Behavioral Dimension. This dimension addresses the principles, policies, and governance rules that drive or constrain the behavior of the CNO and its members over time. Included here are elements such as principles of collaboration and rules of conduct, principles of trust, contracts, conflict resolution policies, etc.

Note that not all generic elements relate to all subcategories, and the detail may be different for strategic and goal-oriented networks.

Moving on to the external viewpoint, three generic elements are considered in each of four subcategories:

- Network identity—defining the general positioning of the CNO in the environment or how it presents itself to the environment.
- Interaction parties—identifying the relevant entities with which the CNO interacts.
- Interactions—listing the various transaction types between the CNO and its interlocutors.

The four subcategories are:

- A Market Dimension. This dimension covers the issues related to the interactions with both "customers" (or potential beneficiaries) and "competitors". The customer facet involves elements such as the transactions and established commitments associated with contracts, marketing, and branding. On the competitor side, issues such as market positioning, market strategy, and policies can be considered. The purpose/mission of the CNO, its value proposition, joint identity, etc., are also part of this dimension.

- A Support Dimension. Under this dimension the issues related to support services provided by third party institutions (outside the CNO) are to be considered. Examples include certification services, auditing, insurance services, training, accounting, and external coaching.
- A Societal Dimension. This dimension captures the issues related to interactions between the CNO and the society in general. Although this perspective can have a very broad scope, the idea is to model the actual or potential impact of the CNO on society (e.g., on employment, economic sustainability of a given region, or potential for attracting new investment) as well as the constraints and facilitating elements (e.g., legal issues, public body decisions, and level of education) the society provides to the development of the CNO.
- A Constituency Dimension. This perspective focuses on the interaction with the universe of potential new members of the CNO, i.e., the interactions with those organizations that are not yet part of the CNO but that the CNO might be interested in attracting. Therefore, general issues like sustainability of the network, attraction factors, what builds or provides a sense of community, or specific aspects such as rules of adhesion and specific "marketing" policies for members, are considered here.

The 64-page document describing the ARCON model included topics to consider within each generic element/subcategory combination, defining each of these lower-level entities. For example, political relations are defined under the interaction element of the societal subcategory that "relates to the interactions with the power groups, often as a lobbying activity, in order to influence political decision making in the domains that affect the CNO". In the same document, a multidimensional interaction matrix is used to map the relative importance of each lower-level element within each evolutionary stage of development, as illustrated in Table 7.4.

7.7 Case Study

This case study could be viewed as illustrating the elements of a system architecture (see Figure 3.1) in a collaboration context where the operating environment may change over time.

Cooperative Research Centers (CRCs) were established under an Australian government program launched in 1990, and supported since

TABLE 7.4

Evolutionary Stage of Development—Multidimensional Interactions

Market	c	o	e	m	d	**Support**	c	o	e	m	d	**Societal**	c	o	e	m	d	**Constituency**	c	o	e	m	d
Network Identity						**Network Identity**						**Network Identity**						**Network Identity**					
Mission						CNO's social nature						Legal status						Attract & recruit. Strat.					
Ref./testimonials												Values & principles											
Network profile																							
Market strategy																							
Interaction parties						**Interaction parties**						**Interaction parties**						**Interaction parties**					
Customers						Certification entities						Governmental organiz.						Business entities					
Competitors						Insurance entities						Associations						Public institutions					
Suppliers						Logistics entities						Interest groups											
						Standard registries						Regulatory bodies											
						Financial entities						Other entities											
						Coaching entities																	
						Training entities																	
						Research entities																	
Interactions						**Interactions**						**Interactions**						**Interactions**					
Advertising						Service acquisition						Political relations						Member searching					
Cust./supp. Trans.						Agreement establ.						Seeking support						Receiving applications					
Handling inquiries												Information transfer											
												Social relations											

Legend:			
Very important		c	creation
Moderately important		o	operation
Not important		e	evolution
		m	metamorphosis
		d	dissolution

then as a long-term strategic initiative. A CRC is a goal-oriented industry/academia/government applied research collaboration funded for around seven years by the Australian government and the partners. CRCs service the manufacturing, information and social services, mining and infrastructure, agriculture, environmental services, and medical service sectors. More than 200 CRCs have been established since 1990. More than 4000 industry-oriented Masters and PhD students have been supported by the program. Long-term studies suggest the program is delivering more than three times the input cost. In 1994, a community of practice, the CRC Association, was established with the intention "to enhance Australia's society, economy and environment through the development of sustained, user-driven, collaborative public-private research centers that achieve high levels of outcomes in adoption and commercialization". The CRC Association can help program aspirants organize new bids, provide a forum for sharing best management practice, and lobby the government if required. It organizes a series of events through the year, including an annual conference.

Submissions for new CRCs or reconfiguration of existing centers are called for every year or two. When proposals are submitted, all parties must have formally committed to provide resources and have commonly agreed to pursue five to seven research themes of national significance. After establishment, the CRC acts like a "breeding network", formulating and approving a succession of projects under each theme. Some CRCs operate in a galaxy configuration with a small administrative core and distributed "stars" hosting different types of projects. Others are established with a large core group and project tasks allocated to "planets". This is common when large-scale research facilities or specialist equipment are only available at one location. Partner organizations host the project groups as part of their contribution to the program.

A CRC is a cooperative entity owned by members from the private or public sectors, together with higher education institutions or affiliated research institutes. A board of governance with directors from (or representing) the members is usually headed by an independent chair, with the CRC's CEO reporting directly to this board. Generally, a technical advisory board reviews and recommends new projects to the governing board. Some CRCs, particularly those with SME participants, have industry association members representing the firms involved, and have found it necessary to establish a broker function to develop project proposals.

Project teams are established in a variety of contexts; a small number of teams undertaking "blue sky" research, some undertaking pre-competitive research with multiple industry partners, and some undertaking applied research with one industry partner. The latter is most common with SME partners who do not have the resources to pick up and progress pre-competitive research within their own organizations.

A 2011 study of the Australian CRC program from a government policy perspective noted changes in the underlying assumptions over time. When the program was first launched it was thought that after an initial seven-year period of funding, the university and industry participants would continue to work together without the need for government funding. From our observations, this is not the norm. The same actors may however join together to seek an extension of CRC funding, or work together on projects attracting funds via other mechanisms. And while firms may adopt collaboration as a strategic tool, other studies suggest they collaborate with customers and suppliers more frequently than with external researchers. Sinnewe et al (2016) studied concerns relating to this continuity aspect in some detail at the level of the participating organizations, adopting a transaction cost theory perspective that examines economic institutions (formal rules and informal customs) and organization transaction methods (how value is exchanged). They suggested that costs associated with collaborative research arose from dealing with three inter-related problems:

- Information—to inform parties about the existence of potential rents.
- Bargaining—to reduce bargaining over distribution of gains.
- Enforcement—to enforce terms of the bargain.

Specific examples were given—firstly, the cost before lodging the CRC application (the information problem); secondly, the costs of negotiations and agenda setting for projects on which the granted funds are to be spent once the CRC application is successful (the bargaining problem); and thirdly, the cost of monitoring and performance evaluation (the enforcement problem).

They proposed further research be conducted in relation to three observations. Firstly, collaboration sustainability bears no relation to the initiating partnership agreement. It is our observation that such agreements are made at a particular time and place, and circumstances may be different seven years later. In addition, some CRCs wind up because they have reached their intended goal. Secondly, collaboration sustainability is supported by strong relationships between members outside of a particular CRC (the extent of social capital). And thirdly, collaboration sustainability is supported by the quality of the CRC experience.

What persists is the enhanced absorptive capacity and social capital developed from working together at a personal level. The authors of this book first met via participation in a CRC project more than 15 years before this book was written, and during that period have worked on a variety of applied research and teaching projects. Drawing on interviews with participants in four CRCs, Hayes (2011) described the evolution of informal "knowledge-stewarding" communities of practice and different temporal perceptions. Knowledge-stewarding communities are focused on organizing, upgrading, and distributing knowledge frequently used by members, and in the CRC context perform a boundary-spanning function. We have also been involved in other CRCs and a variety of other research collaborations.

We also observe that temporal considerations have a significant influence. Academic time clocks are conditioned by the background semester/year rhythm of a university and the three- to four-year cycle time of a PhD study. Industry time clocks are driven by the time windows available to exploit the outcomes of research that sustains a competitive advantage, and by background rhythms associated with technology uptake in an industry sector. These rhythms are short in the IT sector, for example, but long in the pharmaceutical sector. In addition, SMEs tend to focus on shorter time horizons than larger firms. It is our observation that there is better temporal matching within some CRCs than in others.

Viewing CRCs from a researcher perspective has suggested a number of tensions that should be considered:

- How the competing goals and expectations of partners can be integrated within a single collaborative center.
- How trust is (or is not) generated and the implications this has for the way partners interact with and treat each other.
- How the careers of scientists might change through the choices they have within CRCs.

The foregoing discussion is mainly oriented towards the internal workings of CRCs, and could be regarded as reflecting the structural, componential, functional, and behavioral dimensions referred to in section 7.3. We now reflect on the external dimensions of market engagement, external support, societal perspectives, and constituency.

While the participating organizations may be seen as the "customers" of a CRC, each of them has networks in their respective "markets": markets for knowledge assets, markets for research services, and markets for commercial goods and commercial or community services. Over the lifecycle of a CRC, these markets are liable to change, influencing both the nature of projects chosen and attitudes towards CRC participation. We have observed frustration on the part of some industry executives who sought to influence projects, only to be told that funds were already committed to research projects for the life of the CRC. Support for and by a CRC is associated with its network identity—what it was established to research, and how this is communicated externally.

The CRC Association performs an external support function on behalf of CRCs generally, sharing success stories. Functional support to CRCs may be provided via partner channels, (e.g., internet connections as an in-kind contribution) or may have to be provided by independent body (e.g. financial auditing), and both types may be associated with separate long-term agreements. The external societal dimension of CRCs is dominated by government policy deliberations. Over a 20-year time span, an initial focus on addressing the perceived commercial failure of the program due to limited academia–industry research interaction has been supplemented by consideration of other community needs, e.g., as indicated by the funding of a Bushfire CRC [7]. The constituency dimension is associated with the ability to attract new members, and the establishment of mechanisms for receiving and processing new member applications. Arrangements for accepting new members or dealing with departing members are reflected in the constitution of each CRC as a requirement to support government funding. Again, the CRC Association broadcasts new opportunities to participate beyond the networks of current members.

7.8 Reflections

Engineering and operation of systems of systems frequently requires cross-disciplinary knowledge sharing involving human interaction and exchanges of codified knowledge. In addition, individuals may collaborate to share other kinds of resources to achieve a common goal. The orchestration of collaboration activities is facilitated through a combination of communicator, coordinator, and integrator actor functions. Relationships established and operational protocols developed during a collaborative project may endure beyond the lifecycle of the project, building what has been described as social capital. And while the content of this chapter has been presented from a social sciences perspective, the underlying logic can be and is being emulated in the emerging world of intelligent agents. We invite the reader to reflect on the following, drawing on his or her personal experience:

- Can you think of a collaborative venture you have observed?
- How did this venture get started and how was it managed? Were there tensions to be considered?
- How were the communicator, coordinator, and integrator roles enacted?
- What kinds of protocols facilitated operation of the venture?
- What kinds of relationships persisted beyond the life of the venture?

7.9 References

1. Giddens, A. *The Constitution of Society: Outline of the Theory of Structuration*, Berkeley, CA: University of California Press, 1984.
2. Sako, M. "Does Trust Influence Business Performances?" In *Organizational Trust: A Reader*, Kramer, R.M. (Ed.), Oxford, UK: Oxford Publishing, 2006.

3. Booher, D.E., & Innes, J.E. (2002). "Network power in collaborative planning." *Journal of Planning Education and Research*, 21(3), 221–236.
4. Githens, R.P. (2009). "Leadership and power in fostering a collaborative community in a non-profit professional organization." *Systemic Practice and Action Research*, 22(5), 413–429.
5. Raatikainen, R. (1994). "Power or the lack of it in nursing care." *Journal of Advanced Nursing*, 19(3), 424–432.
6. Sridharan, R. & Simatupang, T.M. (2013). "Power and trust in supply chain collaboration." *International Journal of Value Chain Management*, 7(1), 76–96.
7. Sinnewe, E., Charles, M.B., & Keast, R. (2016). "Australia's Cooperative Research Centre Program: A transaction cost theory perspective." *Research Policy*, 45(1), 195–204.
8. McGinn, K.L., & Lingo, E.L. "Power and influence: Achieving your objectives in organizations." Harvard Business School, Background Note 801–425, March 2001. (Revised July 2007.)

7.10 Additional Reading

Camarinha-Matos, L.M. & Afsarmanesh, H. (Eds.) *Collaborative Networks: Reference Modeling*, Springer Science & Business Media, 2008.

8

IT Infrastructure and the Internet of Things

(The internet offers) ... "the broad opportunities for most companies involve supplying information or entertainment. No company is too small to participate."

Bill Gates
Content is King, *January 1996*

8.1 Internet of Things

In today's global economic setting, it is clear that an individual organization alone can no longer prosper in business. Instead, an entire network of firms is involved in the delivery of goods and services to end users [1]. This entire network should be analyzed and managed in order to produce and distribute merchandise in the right quantities, to the right locations, and at the right time. Information is critical to the effective coordination and integration of business partners in a global business network.

Before the advent of the internet, however, linking information systems across the global business network was expensive and technically challenging. As a result, the development of global business networks has been slow, and unaffordable for SMEs.

The proliferation of the internet since the early 2000s has offered promising opportunities to global business networks. However, some significant challenges need to be overcome before these opportunities can be fully realized. Information management is an important challenge. This has been a central concern because of highly vulnerable and heterogeneous information systems among supply chain partners (e.g., causing data incompatibility), which impose unique challenges in terms of scalability, performance and manageability when developing new IT infrastructure incorporating legacy systems [2].

Moving crucial management decisions and processes to the internet also presents serious security and privacy issues. Last but not least, internet-enabled global business networks can not only cause technical challenges, but also social and economic issues. For example, constant education is needed to make staff aware of cyber security threats on the internet.

8.2 Business Systems over the Internet

Businesses are increasingly using the internet for their IT infrastructure. The internet enables optimization of the operation of a global business network by maximizing information sharing, thereby offering significant opportunities for cost reduction and service improvements. For instance, many companies have implemented point-of-sale scanners that record what is being sold in real time. These companies not only collect information to make decisions on what to order or how to replenish their stock, but send this information, through the internet, to their suppliers for synchronization of production with actual sales. This section discusses some of the common internet-based business systems currently in use.

8.2.1 Purchasing/Procurement over the Internet

One of the key processes of global business networks, purchasing the desired product at the lowest price and highest quality, is one of the critical factors in the success of a company. During the past decade, many internet-based purchasing/procurement systems have been developed and used throughout the world. Examples include electronic catalogs, price search engines, recommendation agents, and comparison matrices. Recently, internet technologies have been increasingly used in other aspects of purchasing activities such as customer-need recognition and its technical specification, supplier search, and evaluation of purchase alternatives.

Due to the fact that internet applications in procurement provide a range of solutions that are cheaper and easier to set up than developing special in-house solutions, managing purchasing through the internet has rapidly become widespread. Internet-based purchasing is widely used for a variety of purposes:

- Purchasing and paying for goods and services online. Electronic Data Interchange (EDI) was the earliest e-procurement system, dating to the late 1960s. Many e-procurement systems have been developed and implemented to substantially improve the procurement process since the 1990s. Through the internet, an e-procurement system can provide a customized catalog that provides a rapid response to a search request by customers. Applications that support electronic reverse auctions can enable a company to buy goods or services at the lowest price or a combination of lowest price and other conditions via the internet. The adopters of e-procurement systems worldwide have demonstrated significant improvement of supply chain management.
- Implementing a global procurement strategy. Through internet catalogs provided by vendors, a company can check the vendors'

price quotations online. This gives the company greater choice. Moreover, the company can transcend geographical limitations and purchase from vendor's website at any time, unlike with traditional procurement methods.

- Bargaining and renegotiating price and term agreements. With the support of advanced internet technologies such as online chatting, companies can easily discuss issues related to price, term agreements, warranty, etc., thereby increasing the efficacy and efficiency of procurement.
- Managing and handling product-damage issues. Internet-based purchasing can reduce the costs of handling returned or damaged goods by using fast tracking and notification mechanisms before the damaged goods are shipped. Additionally, the financial aspects of goods return are handled more efficiently through the internet, e.g., for notification of credits posted by vendors. Similarly, warranty issues can be handled over the internet.

8.2.2 Inventory Management

Inventory management is one of the critical activities of supply chains since each one-point drop in inventory cost can result in significant savings, millions of dollars, perhaps, for large companies. Research has shown that a lack of efficient information flow in the management of inventory is a major reason for high costs and inefficiencies in global business systems [3]. Internet technologies are able to provide precise solutions to facilitate the information flow.

Research by Mahar [4] shows that many companies use the internet to manage their inventory successfully. Internet-enabled inventory management systems can provide real-time stock information at all levels in the supply chain from customer stocks, field inventories, and plant inventory, even for stocks and inventories that are geographically distributed. The most popular use of the internet in the inventory management area is the notification of stock-outs by companies to their customers or communication of stock-outs by customers to vendors. Internet-based inventory management has some major advantages:

- Companies can be proactive in managing inventory systems. By using the internet, a company can track out-of-stock items in field depots, notify customers of order-shipping delays, and inventory emergencies. The company can also envisage the demands from the available order information of customers, and take actions to replenish the inventory quickly when it is needed.
- Senior management can get inventory information more quickly and make prompt decisions with internet-based reporting systems.

The inventory information might include finished-goods inventory levels at manufacturing and field depots along with raw material levels at central and regional assembly locations.

- Important items can be tracked in a timely and efficiency manner by integrating near-field communication technologies with the internet [5]. With these technologies, a company can implement online, real-time tracking systems that provide paperless, timely, and high quality information that can maintain inventory control for any specific items, thus improving customer satisfaction and profitable operations.
- Significant stock reduction or even "zero inventory." The idea of this practice is that the retailer serves as a middleman who acquires customers and accepts orders, while the wholesaler owns and stocks the product and ships it directly to customers at the retailer's request [6].

8.2.3 Transportation

Transportation is one of the highest cost components in a supply chain. To reduce this cost, many technologies that support transportation management over the internet have been developed. For example, internet fleet management systems allow customers to track their items at any time around the world. By using these technologies, productivity of transportation including outbound truckload and carload shipments is improved, the back-haul rate is reduced, and claims management is improved, resulting in overall economic gains. For example, a transport business applied an internet-based transportation management system to the forestry industry. The result has shown that the system, which has been integrated with GPS technology, has significantly reduced the waiting times of trucks at the forest and at the destination, thus reducing the overall cost of transportation.

Internet-enabled transportation management systems can provide other functionality:

- Monitor pickups and drop-offs at regional distribution centers by carriers. This is particularly important for a company, since data from the tracking of shipments to regional depots can enable a company to estimate the reliability performance of the carriers it is using. This enables transportation managers to make sure that the carriers they use are meeting their time constraints. It also provides managers with the information needed to inform customers about delays as they occur, and take corrective measures to reduce negative impacts of the delays.
- Manage the claims occurred in the transportation management. Claim tracking through the internet is also possible and can be efficient. Claims reporting, processing, and settlement are more easily handled by an internet tracking system.

8.2.4 Order Processing

One of the benefits of ordering through the internet is that it provides mechanisms to streamline the quotation process and lower the overall cost:

- Order placement and order status checking. Electronic order processing and status checking can reduce the costs of order processing in the supply chain. A major reason of this cost saving is the reduction of paperwork involved in traditional order processing systems.
- Improvement of speed and quality of order processing. Internet-enabled order processing systems can reduce the time between the order being placed and when the product is received by a customer. Moreover, the use of the internet in order processing can reduce the error rate of the traditional paper-based ordering system. This is mainly because the software used in internet order processing systems have standard user interfaces that require less human involvement and paperwork in the ordering process.
- Handling of returned goods and out-of-stock notifications for customers. One of the typical functionalities provided by an internet-enabled order processing system is the provision of timely responses in the ordering processing when the items are out of stock, the wrong items are sent to the customers, or the quality requirements of customers are not satisfied. This function is economically beneficial, since a return order can cost a company as much as 12 times more to process than an outbound order. Thus, better management of this process can help companies reduce the related costs.

8.2.5 Customer Service

The internet offers service improvement opportunities to the organization in addition to improving the selling of its products and reducing the operational cost of its business processes. The ability to provide worldwide customer service, to track systems status of third-party service providers, and to reduce service costs and response time are also widely acknowledged in many engineering-oriented service systems. There are major benefits in using internet-enabled customer service:

- Enabling a company to receive customer complaints and respond quickly. Internet-based customer service systems can provide timely and directed responses to customers located in different areas, even in different time zones.
- Enabling a company to issue emergency notifications when problems occur.

- Providing customers 24-hour access to a company's service department, and allowing customers to immediately notify companies about any service issues or problems that may arise.
- Developing a healthy relationship with customers. Research has shown that customers whose service issues are dealt with quickly and to their satisfaction are more likely to purchase the company's products again [8]. Thus, the internet can help build product and service loyalty and strengthen customer-company relationships.

To maximize customer satisfaction, an automatic customer service system that can automatically handle customer requests by analyzing the contents of requests and finding the most feasible answers from FAQ databases can be used [9]. Such a system can significantly reduce service cost and provide more efficient and effective customer service.

8.2.6 Production Scheduling

The internet can facilitate production scheduling in other ways:

- Coordinating just-in-time (JIT) programs with vendors or suppliers. The JIT system was developed by Toyota in the 1960s. The full potential of JIT is hard to achieve without efficient communication and exchange of information, and the internet provides an efficient platform to include vendors into production scheduling activities. Through the internet, suppliers can receive orders in real time from companies and use the information infrastructure of the entire supply chain to enable JIT production of each specific order.
- Coordinating production schedules between companies and their vendors and suppliers. This can be done both domestically and internationally.
- Conducting customer-demand analysis. The internet has provided a real opportunity for customer-demand data and supply-capacity data to be visible to all companies within a supply chain [10]. Effectively using this information along with other information such as the information derived from internet customer demand surveys will likely produce more accurate forecasts for the changes and fluctuations of markets, thus helping to schedule production more effectively.

8.2.7 Managing Relations with Vendors

As mentioned before, the internet enables vendors and customers of companies to handle various business activities of supply chain on a 24-hour basis, providing an effective mechanism to develop better relationships with customers and vendors. Research has shown that providing vendors with ratings of the on-time performance of their carriers and their overall service performance helps improve the relations between a company and its vendors [11]. The performance rating includes factors such as fulfillment of the deliveries to company warehouses and depots, the on-time performance of the carriers used by the vendors, the quality of the raw material from vendors, and the vendors' raw material inventory and general stock levels.

Internet-based evaluating systems can help to (1) better understand and improve the overall quality of vendor performance, (2) lower purchasing costs, and (3) improve the productivity of vendor operations. Moreover, this evaluation information enables companies to form strategic vendor alliances built on solid informational bases developed from internet monitoring systems. Besides, using internet technologies in supply chain processes likely leads to improvement of the relationships between a company and their customers and business partners. For example, procurement processes through the internet help to improve inter-organizational coordination and cooperation within the supply chain, and offer competitive sourcing opportunities for the buyer organizations.

8.2.8 Engineering and Design of New Products

Complex products such as automobiles, aircraft, and weapon systems require significant engineering and design analysis effort involving a large number of companies, particularly those which are supplying components for the final product. The management of this type of supply chain is complex, time consuming, difficult, and fraught with errors. For example, in the defense business, the engineering trade-off process in the conceptual design phase can take up to a year before all the information requested is available for the design engineering team to start work. A typical problem encountered is the inability of small suppliers far up the chain to access technical data.

The concept of a global concurrent engineering platform using the internet for supporting intercontinental product development team started in the late 1990s. Jiang and Mo [12] developed a distributed design system that could link designers to cowork on the same product over the internet. Designing a new product involves a great deal of analysis and specialized computational tools. The availability of an internet-based collaborative platform brings the most appropriate resources and expertise from different parts of the world to the design team, which can work on the product round the clock, thereby shortening the engineering development time.

Likewise, Mills et al [13] created an information infrastructure, which they called a systems integration architecture, using distributed object computing and remote process invocation methods. The system facilitated access to remote applications and the flow of information up and down the aerospace supply chain in Texas, USA. More recently, Fixson [14] developed a multidimensional framework that enabled comprehensive product architecture assessments within a supply chain. Increasingly heterogeneous markets, together with shorter product lifecycles, forced many companies to simultaneously compete in the three domains of product, process, and supply chain. Dependencies among decisions across these domains made this competitive situation very complex. This framework focused engineering analysis on product dimensions that were critical for given operational strategies.

The supply chain in engineering and design sector is primarily dealing with the synchronization of design effort through exchange of data. The use of a standardized data protocol is critical. ISO 10303 specifies a standard for the exchange of product data (STEP). This standard is now widely adopted by CAD systems as a data exchange protocol. The internet-enabled supply chain in engineering and design can use this standard data protocol to collaborative effectively.

8.2.9 Industry 4.0

The Industry 4.0 initiative has received an increasing amount of attention in recent years. The initiative specifies the needs of a future production environment that allows individual, customer-specific criteria to be included in all phases of the product development lifecycle, and to enable last-minute changes to be incorporated. The concept of Industry 4.0 is based on the evolvement of manufacturing and in-depth customization (i.e., customizing a product with high degree of product configuration). The advancement of the internet, e-commerce, and social networks empowers consumers with more product details, information about new product launches, and in-depth product reviews. It is envisaged that the global manufacturing sector will undergo a transformation to produce highly customized products tailored to individual needs.

In this scenario, the market will demand variable and heterogeneous products. To gain more competitive advantage and to satisfy increasingly sophisticated consumers, manufacturers have to find effective and efficient ways to engage customers and create individually customized products with the shortest lead time. The production strategy of enterprises depends on the manufacturing models which have been evolving continuously in the last two centuries. Several paradigms have appeared.

The first paradigm, craft production, emerged around the mid-18th century. In this paradigm, customers have to bear a high cost for the product they desire and the production is at low efficiency.

The next paradigm, mass production, was then created by the development of moving assembly lines. This paradigm offered low-cost products through large-scale manufacturing producing a very limited variety of products effectively and efficiently. In this paradigm, designers exclusively determine what and how to produce end products without customer involvement.

To cope with the trend of globalization and customer requirements for high product variety, the, mass customization paradigm emerged in the late 1980s. Benefiting from the introducing of modularization, mass customization is an improved solution which combines the low unit cost of mass production processes with the flexibility of individual customization. To cope with personalization requirements, new manufacturing models have been proposed to handle more specific customer needs. OKP (one-of-a-kind production), emphasizing the "market of one," aims to satisfy individual customer requirements while maintaining production efficiency. In this paradigm, the use of internet is exploited to assist information exchange between companies.

Evolving from the three earlier manufacturing paradigms, Industry 4.0 is regarded as a fourth paradigm that will depend on mature use of the Internet of Things (IoT). A variety of information technologies including cloud computing, virtualization, service-oriented technology, and advanced computing technologies has been integrated with existing cutting-edge manufacturing models. The Industry 4.0 paradigm encompasses ubiquitous, convenient, on-demand network access to a shared pool of configurable manufacturing resources (e.g., manufacturing software tools, manufacturing equipment, and manufacturing capabilities) that can be rapidly provisioned and released with minimal management effort or service provider interaction. In some cases, Industry 4.0 is thought of as a computing and service-oriented manufacturing model developed from existing advanced manufacturing models (e.g., application service providers, agile manufacturing, networked manufacturing, and manufacturing grids) and enterprise information technologies under IoT.

Since the main concept of Industry 4.0 is virtualizing of manufacturing resources and capabilities, a rapid response to dynamic demands for product customization can be achieved through the retrieval and consumption of manufacturing services. For the fabrication of complex customizable product that necessitates effective collaboration in the distributed global manufacturing environment, the use of IoT as the backbone will significantly reduce management efforts in the production process. Effectively, Industry 4.0 aims to configure global manufacturing resources dynamically, with a high degree of system configurability for product personalization.

8.3 Challenging Issues in Developing Internet-Based IT Infrastructure

With the advancements in IoT, global business networks rely more and more on the information technology and communication infrastructure to do business. In this section, we explore the issues of changing the business mode from traditional means to an internet-based system. Typical issues in supply chain operations include short product lifecycle, supply variability, collaboration, confidentiality, intellectual property, conflicts, opportunity loss, capacity constraints, and others [15]. The use of information and communication technologies is not enough to foster an efficient flow of information for effective decision-making. A strategic framework is required to integrate the upstream and downstream managers in the supply chain as part of a team that creates and adds value to the products that end up in the hands of the consumer. Therefore, future supply chains have significant challenges as the technology enables the supply chain to do business over the internet. These challenges can be categorized in three major aspects: social, technical, and economical.

8.3.1 Social Issues

A supply chain consists of a number of organizations with different business perspectives (such as suppliers, manufacturers, distributors, and customers). Its effective management requires integration of all parts of the supply chain. Hence, although the use of electronic commerce has been increasing since early 2000s, the rate of adoption in supply chain management has been slower than expected.

A critical issue in the adoption of information technology in a supply chain is the social pressure exerted among the supply chain partners. Managers in the global business network should understand the institutional pressure they were subjected to when working with their supply chain partners. They might face, in the course of adopting IT for the management of their supply chains, possible problems and restrictions. Different types of institutional isomorphism, namely coercion, mimesis, and norms, could impose discrepancies from deteriorating supply chain performance to losing business opportunities.

Therefore, a crucial task in building a supply chain is to clearly define the social context in which the supply chain is going to operate. The organizations that are potentially involved in the supply chain, whether willingly or unwillingly, need to spend time to determine the physical and structural properties of collaboration, the culture and business practices, and the security processes and governance issues. Possible problems in the supply chain, e.g., conflict of interests, confidentiality, and intellectual property ownership, should be identified well beforehand.

8.3.2 Technical Issues

An internet-enabled global business network requires product information to be transferable in electronic format. A crucial condition to enable this capability is the compatibility of product lifecycle information models for decision-making based on data gathered through different parts of the product development, manufacturing, sales, and services processes [16].

The fundamental research required is in information system models, smart embedded systems, short and long distance wireless communication technologies, data management and modeling, design for X and adaptive production management for beginning-of-service life of products, statistical methods for predictive maintenance, and planning and management of product end-of-life.

While the trend is clear for companies to make best use of IoT as the way forward, much work is still to be done on understanding and modeling the changes to internet trading and their effect on managers. These include but are not limited to these issues:

- **Global business network visibility and efficient tracking**—Data transparency requires enforcement of data flow and contextual integrity in the supply chain. It is important that the corresponding business processes among the trading partners are developed and synchronized to ensure a high visibility of information flow throughout the system.
- **Network performance management**—Unlike a normal enterprise, the supply chain is formed from a number of autonomous enterprises. Individual enterprises may choose to work in a certain way to optimize their own performance but this may not optimize the ultimate performance of the supply chain. It is therefore necessary to develop key performance indicators that can reflect whether the goal of the entire supply chain is met. However, partnering companies have their own practices and standard procedures when dealing with one another in various business matters. For example, there are many different file formats and data structures used for scheduling and costing. The incompatibility of data and systems among the supply chain partners make it difficult to determine the overall performance of the whole supply chain.
- **Security and privacy**—For each business process there are a number of headline systems that it relies upon. These interrelationships should be mapped out in a process known as situational awareness. By accurately mapping business processes to information systems, system operational risk can be identified leading to real business risk and subsequently vulnerability could be determined from analyzing the graphical representation of the system.
- **Dynamic and scalable processes integration**—The global environment is dynamic and often affected by customer preferences such as seasonal requirements. Four risk-influencing determinants—forecast uncertainty, demand variability, contribution margin, and time window of delivery—contribute to the responsiveness of the toy supply chain. In addition to the dynamic characteristics of demand, the global market is changing to a customer-dominant environment.

Many customers expect a build-to-order strategy to be adopted to meet the requirements of individual customers by leveraging the advantages of outsourcing and IoT.

8.3.3 Mass Customization

Customers demand more individualized design and products, which forces changes upstream of the supply chain not only in satisfying quantity but also in meeting variety expectation. The effect is the need for a scalable manufacturing system that can integrate large number of design alternatives with flexible manufacturing facilities while managing fluctuation in production volume.

Mass customization is the key to satisfy this demand. Mass customization production is a challenge to existing production management systems. It would be difficult to maintain efficient use of the production system for the build-to-order approach which is most often associated with a customization strategy. Each item in the global business network will have to be individually identified to enable customization of features and functions to meet customer needs. There are two aspects of the customization process. First, the customer should be given the opportunity to specify his or her needs. Second, the supply chain should have built-in functionality that can respond to individual customer specifications. Both aspects affect the effectiveness of production planning that depends heavily on availability of information from IoT.

It is clear that a strategy based on reliable information infrastructure for supporting dynamic supply chains is important. In this scenario, an internet-enabled supply chain will be able to facilitate better information flow among the collaborating partners. Decisions on the best production routes within the supply chain to meet the need of particular customer order could be made and supported by reliable data. The ability to call on resources from anywhere in the world is vital to meeting the scale and variability of customer demand.

8.4 Economic Issues

A supply chain is a loosely coupled entity that is driven by economic factors; that is, to share the benefits of trading to fulfill customer demand. Likewise, risks can be mitigated by sharing market information across the supply chain so that member companies can consider risk management opportunities to hedge against aggregate economic risk. Problems such as partner selection, operation management, information exchanges and application of different standards could affect the economic viability of the supply chain [17]. However, they are difficult to quantify and evaluation of risk requires more sophisticated analytical methodology.

There is a considerable gap in understanding of the bottom-line implications of supply chain strategies, in which e-commerce scenarios are playing an ever greater role. Using three scenarios (traditional, electronic-point-of-sale-enabled, and vendor-managed inventory), Naim [18] investigated the impact of a production planning and control system. The study used the net present value (NPV) to show that management decisions are sensitive to the selection of parameters in the ordering process. Naim's work has been useful for further research on costing supply chain dynamics.

Ettlie et al [19] showed that there were obstacles to investing new internet-based technologies in companies which had prior investments in earlier generations of e-commerce technology such as EDI. In fact, they identified that leadership (social aspects), business process engineering (organizational aspects), and acquisition strategy (business culture aspects) were important factors in IT implementations. In the retail supply chain, large retailers could benefit from high IT investment on several fronts. In this case, the consideration for investment in new IT systems did not rely solely on return. The strategies suggested were to use their enormous financial power to drive technological advancement to their advantage as well as to pass on the technology costs to their vendors.

Kärkkäinen and Holmström [20] believed one of the greatest challenges for the adoption of advanced supply chain technologies such as the internet and RFID was the lack of investments in infrastructure. There were always disagreements about the sharing of investments in the technology and the benefits obtained. Business investors are unsure about return on investment. Even if they have some indicative information, the returns so far are not attractive. However, to achieve reasonable benefits from internet-enabled supply chain technology, massive participation is required. The challenge for the entrepreneurial supply chains is to increase community confidence to generate a larger scale of participation.

8.5 Reflections

From your experience, identify a case in which at least 80% of the supply chain transactions are completed over the internet. Develop a process description (this could be done with a process modeling tool or just by listing the activities step by step) of the transactions in the case and indicate clearly which activities are internet-based and which are not. Estimate the total transaction time.

Now wind back in time and explain how it was done before the availability of the internet. Develop a new process description (using the same format and notation) of the transactions in the wind-back case. Estimate the wind-back transaction time.

8.6 References

1. Angeles, R. (2005). "RFID technologies: supply-chain applications and implementation issues." *Information Systems Management*, 22(1), 51–65.
2. Lancioni, R.A., Smith, M.F., & Schau, H.J. (2003). "Strategic Internet application trends in supply chain management." *Industrial Marketing Management*, 32(3), 211–217.
3. Fleisch, E. & Tellkamp, C. (2005). "Inventory inaccuracy and supply chain performance: a simulation study of a retail supply chain." *International Journal Production Economics*, 95(3), 373–385.
4. Mahar, S. (2009). "The value of virtual pooling in dual sales channel supply chains." *European Journal of Operational Research*, 192(2), 561–575.
5. Boone, T. & Ganeshan, R. (2007). "The frontiers of eBusiness technology and supply chains." *Journal of Operations Management*, 25(6), 1195–1198.
6. Ayanso, A., Diaby, M., & Nair, S. (2006). "Inventory rationing via drop-shipping in Internet retailing: a sensitivity analysis." *European Journal of Operational Research*, 171(1), 135–152.
7. Sikanen, L., Asikaninen, A., & Lehikoinen, M. (2005). "Transport control of forest fuels by fleet manager, mobile terminals and GPS." *Biomass & Bioenergy*, 28(2), 183–191.
8. Levenburg, N.M. (2005). "Delivering customer value online: an analysis of practices, applications, and performance." *Journal of Retailing and Consumer Services*, 12, 319–331.
9. Tseng, J.C.R. & Hwang, G.J. (2007). "Development of an automatic customer service system on the internet." *Electronic Commerce Research and Applications*, 6(1), 19–28.
10. Kehoe, D. & Boughton, N. (2001). "Internet based supply chain management: a classification of approaches to manufacturing planning and control." *International Journal of Operations & Production Management*, 21(4), 516–525.
11. Opoku, RA (2007). Electronic supply chain management applications by Swedish SMEs. *Enterprise Information Systems*, 1(2): 255–268.
12. Jiang H.C., Mo J.P.T. (2001). Internet Based Design System for Globally Distributed Concurrent Engineering. *Journal of Cybernetics and Systems*, October-November, 32(7):737–754.
13. Mills, J., Brand, M., Elmasri, R. (1999). "AeroWEB: An Information Infrastructure for the Supply Chain", in *Information Infrastructure Systems for Manufacturing II*, ed. Mills, J.J., Kimura, F., pub. Kluwer, ISBN 0-412-84450-8, Ch.22, pp. 323–336.
14. Fixson, S.K. (2005). Product architecture assessment: a tool to link product, process, and supply chain design decisions. *Journal of Operations Management*, 23:345-369
15. Sheffi, Y., Rice, J.B.Jr. (2005). "A supply chain view of the resilient enterprise." *MIT Sloan Management Review*, 47(1), 41–48.
16. Mo, J.P.T. & Zhou, M. (2003). "Tools and methods for managing intangible assets." *Computers in Industry*, 51(2), 197–210.
17. Li Y. & Liao, X. (2007). "Decision support for risk analysis on dynamic alliance." *Decision Support Systems*, 42, 2043–2059.

18. Naim, M.M. (2006). The impact of the net present value on the assessment of the dynamic performance of e-commerce enabled supply chains. *International Journal of Production Economic*, 104(2):382–393.
19. Ettlie, J.E., Perotti, V.J., Joseph, D.A., & Cotteleer, M.J. (2005). "Strategic predictors of successful enterprise system deployment." *International Journal of Operations and Production Management*, 25(9-10), 953–972.
20. Kärkkäinen, M. & Holmström, J. (2002). "Wireless product identification: enabler for handling efficiency, customization and information sharing." *Supply Chain Management: An International Journal*, 7(4), 242–252.

9

Governance

> No matter how an individual views Satan, whether they believe that he is a real character or that he is just the product of literary scholars and imaginations, no one can deny that each one of us has an aspect of the devil within us. By studying the character and nature of Satan, we learn about ourselves; and the more we know about ourselves, the better we can fight our own personal demons—metaphorical or otherwise—in order to create a better tomorrow.
>
> **Nwaocha Ogechukwu**
> *The Secret Behind the Cross and Crucifix, 2009*

9.1 What is Governance?

Governance is the action to comply. One of the key stakeholders in a system of systems is the regulator. A regulator is often a government agency set up to monitor and control the conduct of business entities, and is not primarily interested in cost modeling, but can have a significant impact on a system of systems. To appreciate the regulator's role and impact on through-life service and support it is first necessary to understand the relevant regulatory systems governing a particular industry. Most organizations have regulatory stakeholder engagement and feedback mechanisms in place. It is advantageous to enact and promote an environment of continuous improvement with the relevant regulatory bodies.

There are a number of prerequisites for a regulatory system:

- Harmonized policies
- Organizational structures that define the independent regulatory body
- Processes for assurance of compliance with those regulations
- Feedback systems to prevent over-regulation, i.e., to mitigate the implications of compliance

Regulations are promulgated for the assurance of business conditions that are acceptable to the community. Governance in systems of systems aims to comply with regulations. For example, the regulations for safety of platforms such as aircraft are derived from the public perception of air transport based on over 100 years of knowledge of accidents and consequent recommendations. In most countries, there are many regulators and a broad range of varied stakeholders across a variety of domains, and they may invoke or support international regulatory requirement though bilateral agreements or acceptance of type or class certifications issued elsewhere.

The legislatory system is illustrated in Figure 9.1. At the very top of the pyramid are the acts, which include acts from both federal and state governments. Below the acts are the regulations (state and federal). Underneath the regulations are codes of practice, standards, and guidance notes from government agencies and members of the industry.

Compliance with acts and regulations is mandatory. Failure to comply will result in criminal proceedings. Codes of practice, standards, and guidance notes are advisory, and following them is recommended as "best practice." However, any guidance material, national or international, that is referenced in an act or regulation carries the same weight as the act or regulation, and is therefore mandatory.

All products, including services that are provided together with the product, or are used to operate or maintain the product, must comply with the regulatory framework governing the product within the regulatory regime.

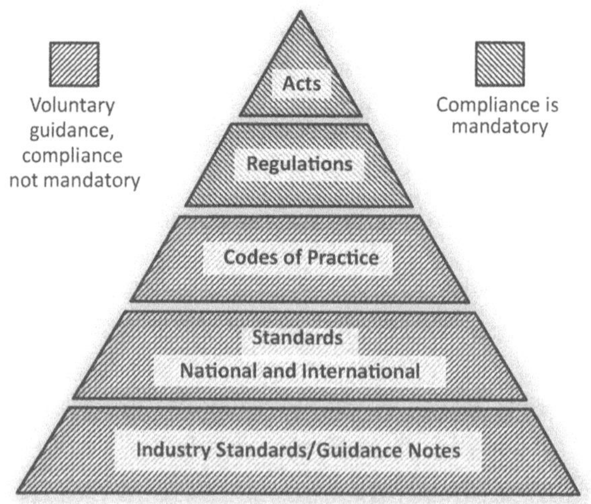

FIGURE 9.1
The pyramid of a regulatory system

9.2 Governance for Enterprises

The first comprehensive governance for commercial enterprises was probably the Sarbanes-Oxley Act, enacted after the Enron scandal. Enron was an oil and gas company from Houston, Texas, which expanded into other industries such as oil refining, power production, pulp and paper, and communications over several decades. In particular, Enron was a pioneer in trading of commodities over the internet. However, its rapid expansion was hard-hit by the bursting of dot-com bubble. The company declared bankruptcy on December 2, 2001.

Enron's collapse was the result of inappropriate accounting practices, dishonesty, and fraud. Its massive loss was reflected by its share price falling from a historic high of US$90.56 to US$0.26 in just one year when it filed for bankruptcy. Subsequently, the US Congress set up a working committee to develop what is now known as the Sarbanes-Oxley Act.

This legislation came into force in the United States in 2002 and introduced major changes to the regulation of financial practice and corporate governance. The act was named after Senator Paul Sarbanes and Representative Michael Oxley, who were its main architects. The Sarbanes-Oxley Act is arranged into 11 titles:

I. Public Company Accounting Oversight Board
II. Auditor Independence
III. Corporate Responsibility
IV. Enhanced Financial Disclosures
V. Analyst Conflicts of Interest
VI. Commission Resources and Authority
VII. Studies and Reports
VIII. Corporate and Criminal Fraud Accountability
IX. White Collar Crime Penalty Enhancements
X. Corporate Tax Returns
XI. Corporate Fraud and Accountability

As far as compliance is concerned, the most important sections are in titles III, IV, VIII, and IX. Section 404 in title IV calls for companies to report on their internal control over financial reporting and requires auditors to render an opinion on that report and the effectiveness of the internal control. In the past, investors and analysts have never been too concerned with the issue of internal control, primarily because management seldom discussed it in their annual reports. In addition, few cared about the adequacy of internal control until the seismic accounting scandals exposed by the Enron case. It was then clear that companies should have proper governance for internal financial controls.

The fundamental intention of the Sarbanes-Oxley Act is to protect investors by improving the accuracy and reliability of company accounting practices:

- Closing loopholes in accounting practices.
- Strengthening corporate governance rules.
- Increasing accountability and disclosure requirements of corporations, corporate executives, and a corporation's public accountants.
- Increasing requirements for corporate transparency in reporting to shareholders and describing financial transactions.
- Strengthening whistleblower protections and compliance monitoring.
- Increasing penalties for corporate and executive malfeasance.
- Creation of the regulator, an oversight board to monitor corporate behavior.

The act also covers financial auditors, who should assess the financial situation of a company independently so that the information reported in the end-of-year reports is checked and verified. Companies are also required to provide, in addition to financial results, evidence of internal controls they have in place to prevent foreseeable problems as well as an assessment of the effectiveness of those internal controls.

9.3 Occupational Health and Safety Requirements

Business corporations, irrespective of their size, should recognize the importance of systems and facilities to support and sustain its operations and developments. The duty of a corporation is to provide a healthy, safe, and secure environment; one which enhances the experience for its customers and allows staff in academic and service areas to work actively and creatively without risk of injury or illness. The first step for achieving this goal is to set up an occupational health and safety policy. This has some important objectives:

- Prevent injury or illness in the workplace.
- Ensure compliance with regulatory requirements.
- Provide a crime-free environment where employees and visitors can work and meet without threat or fear.
- Continually improve the standard of health, safety, and personal security within the workplace.

- Do everything that is reasonably practicable to protect the physical property both of the business entity and of staff, customers, contractors, and all visitors who stay with the corporation temporarily.
- Integrate health, safety, and security into the company's management structures, systems, and strategies.

This is particularly important for service organizations that provide maintenance and enhancement services to assets and systems.

9.3.1 Asset Owner

The asset owner has defined responsibilities as the registered owner of the plant, which carries the generic meaning of any equipment, product, installation, or system.

An owner of a registrable plant must not use it, or allow it to be used, unless the plant is registered. The owner of registrable plant may apply to the appropriate regulatory agency to register registrable plant. It is the responsibility of the asset owner to make sure that registration is valid for the time in which the asset is going to be operated.

A holder of a certificate of registration of a registered plant must give the regulatory agency notice of a change of ownership of the plant in the approved form within a defined period (by the regulatory agency) of the change.

9.3.2 Asset Operator

The asset operator must not install or use a plant unless a certificate of registration of a registrable plant design has been granted for the design of the plant that is in force. The term in force means that the design of the plant is valid at the time when the plant was installed. Changes to the registrable plant design afterward will be applicable to new plants installed but may not necessarily affect the plants which have been installed, i.e. the existing plants are still using the original registrable plant design while the new plant will use the new registrable plant design. If the change is critical, e.g. due to safety, the new registrable design can be reinforced and the existing plants will be required to make the changes in an allowed time. Likewise, an asset operator who is an employer must not allow a worker to install or use the plant unless a certificate of registration of a registrable plant design has been granted for the design of the plant that is in force.

The owner of high-risk plant must not install or use, or allow anyone else to install or use, the plant unless a certificate of registration of a registrable plant design has been granted for the design of the plant that is in force. The certificate of registration of a registrable plant design for the design of plant will be invalid if the design is changed in a way that requires new measures to control risk.

For example, a certificate of design of a mobile crane is invalid if the crane's reach is increased by fitting a longer boom at the end. The length increase has the effect of increasing the risk of the crane overturning. Likewise, a certificate of design of a fire tube boiler remains valid if the change of the boiler is made at the output valve that increases its firing rate, but the operating pressure and temperature in the boiler remain unchanged. This change does not increase the risk of rupture of the boiler since the internal operating stress has not changed.

The holder of a certificate of registration of a registrable plant design for the design of plant must provide all relevant information about the plant (e.g., manufacturer information) to the regulatory agency for inspection and verification, if requested.

9.3.3 Asset Maintainer

The asset maintainer is a person whose work is primarily maintaining, servicing, and/or repairing the plant, e.g., a person who repairs the boom of a crane is not the person who lifts a load with the crane.

The asset maintainer performs work on the plant but does not use the plant for the purpose that it is designed. To ensure that the asset maintainer is a competent person, approval is required based on the conditions that the person

- has a sound knowledge of relevant standards in the country that the asset is operated.
- has a sound knowledge of, and competence in, the risk management process for the erection, operation, maintenance, repair, alteration, and dismantling of the asset.
- has acquired, through training, qualifications, or experience, the necessary skills to design procedures for the inspection, maintenance, and repair of the asset.

9.3.4 Maintenance Support Networks

Provision of successful maintenance support relies on setting up an effective network that includes the supply chain, competent technical personnel, and resources for carrying out maintenance works.

Specialized maintenance support networks have access to technical assistance from experts in the field. In some cases, a global assistance center with professional service resources in the country of the original equipment manufacturer can be organized on international communication networks. Engaging a specialized global services and support organization allows 24/7 services and support to be provided.

This practice is particularly common in network services, where connectivity is provided by specialized telecommunication services. Maintainers can solve problems as well as examine performance of the system being maintained through a special user interface accessing information on the system.

9.4 Systems and Subsystems Design Authorities

Systems and subsystems design authorities have defined responsibilities via regulatory certification for the management of the "Type or Class of platform" of the system. This regulatory requirement is especially well-defined in the aviation industry.

Under a regulatory regime, engineering, maintenance, and operations of an asset require applications for approval and certification of the industry organizations who wish to carry out these activities. The supporting management systems and organizations carrying out these activities are audited by the regulator at regular intervals to ensure compliance to the regulated activities. Regulators often audit on a risk-based approach where they monitor the industry organizations based on the risk profile of the activity (e.g., regular airline public transport activities) and the organizational risk (key people, skills vacancies, organizational change instability, competency, and errors and non-conformances).

Industries working in the regulatory environment are required to have a basic ISO 9001:2008 quality accreditation. Under this accreditation, organizations are required to undertake self-audits of their management systems and facilitate calendar or needs-based external third-party audits. These needs usually arise from applications to change scope or authority in the regulator regime or due to an incident, non-conforming activity being reported, or a reportable error.

To assist the industry, regulators often hold regular forums, symposiums, or conferences as part of their obligation to promulgate regulations, provide training, explain how compliance assurance with regulations is to be assed, and to get feedback to prevent over-regulation, i.e., mitigate the implications of compliance.

9.5 System Safety

Safety means being able to be free from unacceptable risk of physical injury or of damage to the health of people, either directly or indirectly as a result of damage to property or to the environment. This concept can be extended to consider the safety of a system. The primary concern of system safety is the management of hazards as opposed to an emphasis on eliminating component failures in reliability engineering. System safety means the application of engineering management principles, criteria and techniques to optimize the safety of a system, within the constraints of operational effectiveness, time, and cost throughout all phases of the life cycle.

9.5.1 Evolution of Regulations and Standards in Safety

System safety covers a broad range of topics, including the nature of risk, causes of accidents, fundamental concepts of system safety engineering, legislations, and regulations on safety, hazard identification and risk analysis, safe system design, safety analysis techniques, safe software engineering, creating a safety culture, safety cases, safety management systems, human factors, accident investigations and incident reporting, and organizational accident theory.

Regulations and standards continue to evolve as the community or specific regulatory domain discovers deficiencies in existing regulations via incidents accidents or failures of products or processes. New technologies or developments are often not covered by existing regulations or design standards. Regulations and standards governing safe use of new technologies and developments are often put in place when a major incident that no one foresaw exposes the risk. Very often, this regulation is set up too late.

To develop the necessary regulations and/or standards for new technologies or designs, system safety studies must include safe engineering awareness and practices in the areas of engineering design, safety of systems, systems integration safety, and through-life support and safety. Training for situational awareness is becoming popular in many critical operations, and the need for designing safety into new system (and hence system of systems) is absolutely essential.

9.5.2 Acceptable Levels of Safety

The "as low as reasonably practicable" (ALARP) principle is self-explanatory. The implementation of ALARP requires some evaluation (qualitative or quantitative) of the reduction in risk associated with adopting some particular measure, and a clear view of the costs. In some circumstances, both safety risk and the marginal cost or efforts to improve safety can be realistically assessed in numerical terms; in others, risk reduction can only be judged qualitatively—for example, the simple addition of a further safety feature, which costs relatively little, is obviously worthwhile. The principle that safety should be improved beyond the baseline criteria so far as is reasonably practicable is used in the application the primary UK occupational safety act, commonly known as Health and Safety at Work. A risk is considered to be ALARP when it has been demonstrated that the cost of any further risk reduction, including the loss of defense capability as well as financial or other resource costs, is grossly disproportionate to the benefit obtained from that risk reduction.

9.6 Reflections

In this chapter, the Enron case has been used to illustrate the importance of governance compliance and a regulator role. It is not common to personally encounter a governance compliance case. In any event, most of the information about a case remains confidential for a period. However, many cases regarding the result of breaches of corporate governance requirements can be found on the internet.

For a particular case, try to find, on the internet, as much background information as possible and identify which parts of the case have been (or have not been) covered by the Enron case.

9.7 Additional Reading

Dowling, G. (2006). "Reputation risk: it is the board's ultimate responsibility." *Journal of Business Strategy*, 27(2), 59–68.

Part 3

Organizing in Business Networks

10

Virtual Enterprise Collaboration Concepts

> The availability of effective global communication facilities in the last decade has changed the business goals of many manufacturing enterprises. They need to remain competitive by developing products and processes which are specific to individual requirements, completely packaged and manufactured globally. Networks of enterprises are formed to operate across time and space with world-wide distributed functions such as manufacturing, sales, customer support, engineering, quality assurance, and supply chain management.
>
> **John Mo and Laszlo Nemes**
> *Global Engineering, Manufacturing and Enterprise Networks, 2000*

10.1 What Is a Virtual Enterprise?

The developments in global communication networks have significantly changed the way companies operate. Many manufacturing, industrial, service, and commercial activities are now organized into collaborative teams in networked organizations. The operating conditions of any system (irrespective of whether it is the company itself or an in-service engineering system) change frequently due to variations in products, services, processes, organizations, markets, and supply and distribution networks. Autonomous teams often form temporary alliances to deliver a project or product which dissolve when the job is completed. These teams work together as an entity for a goal but maintaining the working relationships among themselves and within their individual companies often relies on trust between individuals and between companies. This type of temporary alliance is commonly known as a virtual enterprise. Success for achieving the goal demands well-coordinated agility in all internal and external aspects of a virtual enterprise.

As the technology for setting up information systems becomes more complex and capable, many companies have adopted the business model the virtual enterprise to collaborate with business partners for the management of the complete lifecycle of system development for customers. Formation of this type of temporary alliance needs to take into account the loss of existing business due to diversification of resources and the loss of opportunity for new businesses.

The business world needs to have the capability to set up virtual enterprises in a short timeframe without making costly mistakes. The virtual enterprise reference architecture and methodology (VERAM) framework was created by a global research project known as Globemen [1] as a way to address this need. Figure 10.1 shows different parts of the research project investigating sections of the cascaded enterprise lifecycles. Over 20 organizations worldwide participated in this project and their interests are shown by the labeled boxes and description of the respective research interests, along with the company logos.

The fundamental concept of VERAM is to set up standard interfaces among participating companies at different stages of the formation of the final virtual enterprise so that companies can adjust their internal processes to suit. At the left of the diagram are the three enterprise lifecycles. The left-most cycle describes the life of the formation of a network of companies with complementary interests and expertise. When the network reaches the implementation phase, some companies agree to collaborate on an opportunity and form a virtual enterprise with a set of defined goals but without entering into any binding contractual obligation. The companies will allocate resources from their own organizations to work with partners collaboratively. At this stage, there could be some promises of revenue but the whole arrangement is based on trust.

The next cycle is the virtual enterprise's own lifecycle until it reaches the implementation phase, where the companies actually receive a firm contract to build a system. Once the contract is signed, the virtual enterprise cycle is transformed into the product lifecycle, shown on the right-hand side. The product lifecycle is essentially the left arm of the systems engineering V-model lifecycle discussed in chapter 2.

10.2 Policies for a Virtual Enterprise

In order to unify the activities of partners in the virtual enterprise and simplify the synchronization work that needs to be done, information technology policies are formulated to provide guidance for the design, development, and implementation of the information infrastructure.

10.2.1 Information Policy

In a virtual enterprise, organizations share investigative experiences in information sharing. They discuss issues related to organization, access, control, and compatibility of information within the context of operations in the extended enterprise.

The information technology infrastructure is designed to support the operation of a virtual enterprise. The virtual enterprise information technology

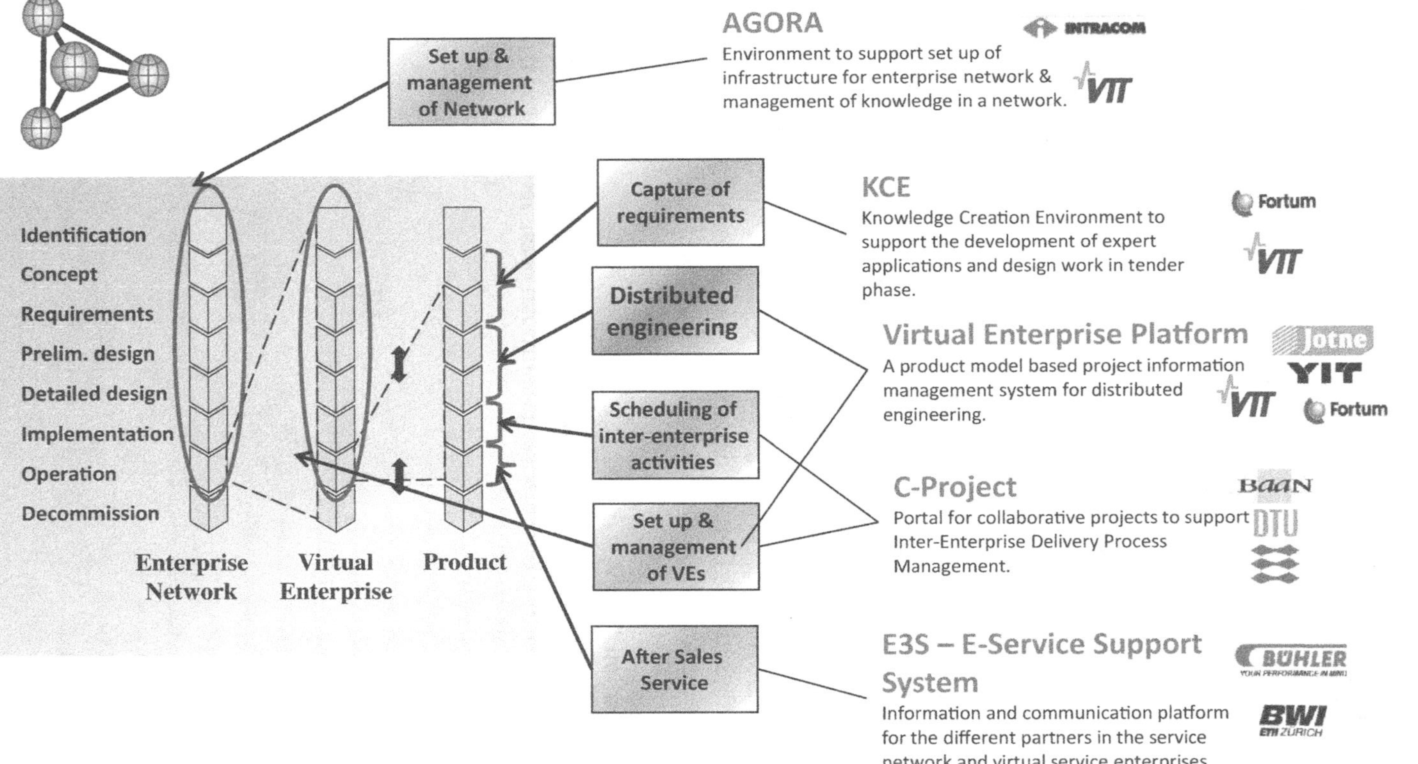

FIGURE 10.1
VERAM research project

infrastructure (VEITI) is developed for one-of-a-kind products designed by a group of companies loosely associated with one another in the global business environment irrespective of the geographical and cultural barriers.

In a single organization, significant resources are devoted to the management of information technology in order to create positive and effective conditions of operation. For the virtual enterprise, this provision is even more crucial to the success of the group. With VEITI, product modeling standards, enterprise reference architecture, and systems engineering principles can be readily applied. These methodologies are integrated with internal IT systems to pull together accurate information and knowledge in a very short period for various business tasks.

The intelligence of the processes in the virtual enterprise requires strong support by the appropriate information technologies. In developing VEITI, reference models and prototypes are proposed as the results of analyses and trials. Considering the rapid evolution of information technologies, this development needs to be fairly swift. The approach adopted therefore aims to be evolutionary, bringing useful results together in a step-by-step manner with existing systems.

The information exchanged on the communication media needs to be secure, authenticated, and use of minimal distributed databases. Communications should be guarded by encryption, by means such as public or secret key encryption algorithms. Digital signatures based on encryption protect against spoofing and illegal modification of the data.

10.2.2 Operation Policy

The widespread application of the internet has enabled a low cost and effective way of transporting information globally. An open network can be constructed as shown in Figure 10.2.

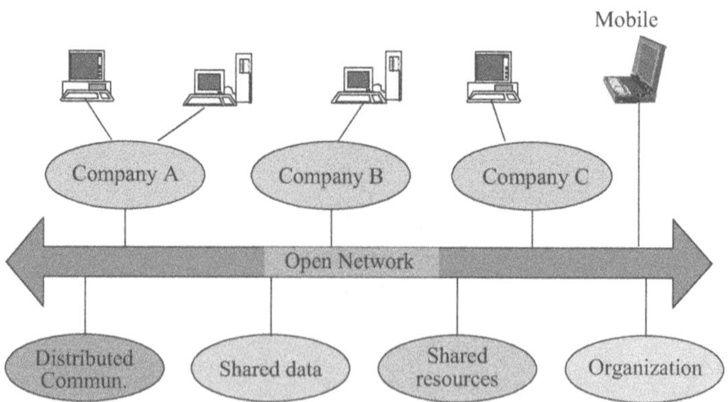

FIGURE 10.2
Open network

By maintaining an open network, clients can enter the system at any time and at multiple points of the plant design. Operations are distributed and the work environment is collaborative and remote.

10.2.3 Design Policy

The information infrastructure should use object oriented technologies, and version control and dependent information must be maintained. Use of multimedia streaming and specific semantics or content-based interpretation embedded is essential in a global information management system.

In addition, the way information is presented to users should enhance human communication and interaction. The goal here is to develop multimedia interfaces that combine voice, video, graphics, and other formats, and display the data to an operator in the most effective manner.

10.2.4 Technology Policy

As far as possible, standards such as STEP (section 8.2.8) and related tools are used throughout the virtual enterprise.

Commercial computer-supported collaborative work systems can minimize development costs. Such tools include video conferencing tools with audio and whiteboards on the communication device. Video conferencing systems are easy to operate. However, they require substantial network resources and real-time transfer. Functionality to record and replay on such tools is useful.

Another useful solution is a group annotation system that supports annotations to documents anywhere on a network shared by the people in the virtual enterprise. Many synchronous and asynchronous collaboration tools are emerging, and these can be evaluated to see how they can be applied most effectively and provide innovation in business processes.

10.2.5 Security Policy

All operations on a VEITI are performed in the presence of company firewalls. A firewall protects the network from invasion, controls access to data determined by user level, and enhances reliability of the network and servers. A typical arrangement of firewalls is shown in Figure 10.3.

The use of firewall presents difficulties for the transfer of information. Packets, irrespective of whether they are useful or not, are not allowed to pass through the organization's firewall if they come from unknown sources. In an extended enterprise situation, this means that new companies may consider there are hurdles to overcome when they join the virtual enterprise. Maintaining the requirements of both security and openness is a challenge.

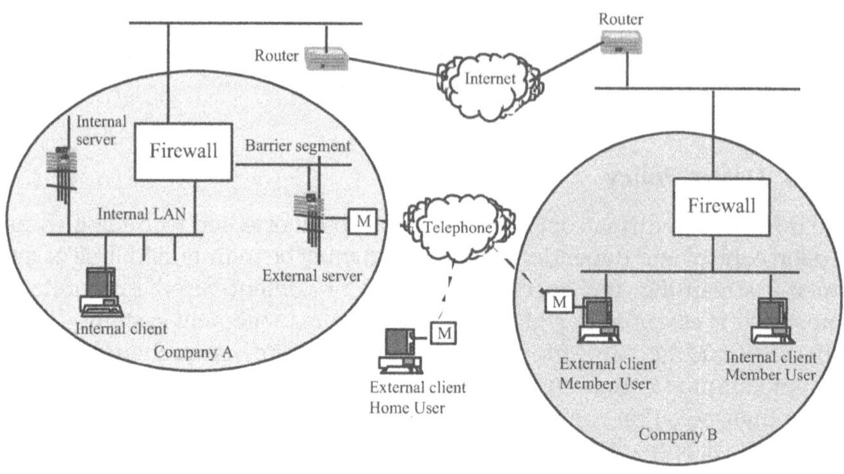

FIGURE 10.3
Firewalls integrated with public and private networks

10.3 A Ship Service Virtual Enterprise

The ANZAC ship is a class of frigates built entirely in Australia. The long service life of a naval ship demands that the asset is adaptable to changing operational requirements. The process of implementing changes on ships was unsatisfactory, with different aspects of change being implemented through multiple contracts. These contracts were often not coordinated, resulting in changes that were too slow, provided limited options, and suffered "sticker shock" within the government department managing these changes [2].

Although the creation of a system program office (SPO) structure for this class of ship lessened the imperative for change implementation, the line of responsibility between platform and combat systems contracts was difficult to define and hard to synchronize. Hence, the Australian government promoted the concept of "alliance contracting" with the expectation that industry collectively could do better.

The ANZAC Ship Alliance (ASA) could be thought of as a virtual company with shareholders comprising the Australian government and two commercial companies, one of which was the ship builder. The primary goal of ASA was to manage all change and upgrades to ANZAC class ships. It was a "solution focused" company where the staff of the ASA management office would develop change solutions but the detailed design would be undertaken by the "shareholders" drawing upon their existing and substantial knowledge of the ANZAC class. The aim was to create mutually beneficial relationships between all parties involved so as to produce outstanding project outcomes. The business model of the ASA was based on these principles:

- All parties win or all parties lose.
- Collective responsibility and equitable sharing of risk and reward.
- All decisions are based on "best for project" philosophy.
- Clear responsibilities within a no-blame culture.
- Access to resources, skills, and expertise of all parties.
- All financial transactions are fully open book.
- Encouragement of innovative thinking aiming at outstanding outcomes.
- Open and honest communication among all partners, i.e., no hidden agendas.
- Visible and unconditional support from executive management.

In order to realize the ASA concept, a tasking process was set up as shown in Figure 10.4. The ANZAC SPO Director, who reports to the Director General in Department of Defence, had SPO managers responsible for different types of change projects. When the need for change arose, the SPO Director issued a tasking statement to ASA general manager who was responsible for carrying out the investigation, while a copy of the statement was sent to the ASA board for record.

The ASA had a small management office to manage the task investigation process. There were no permanent staff. Instead, all staff, including

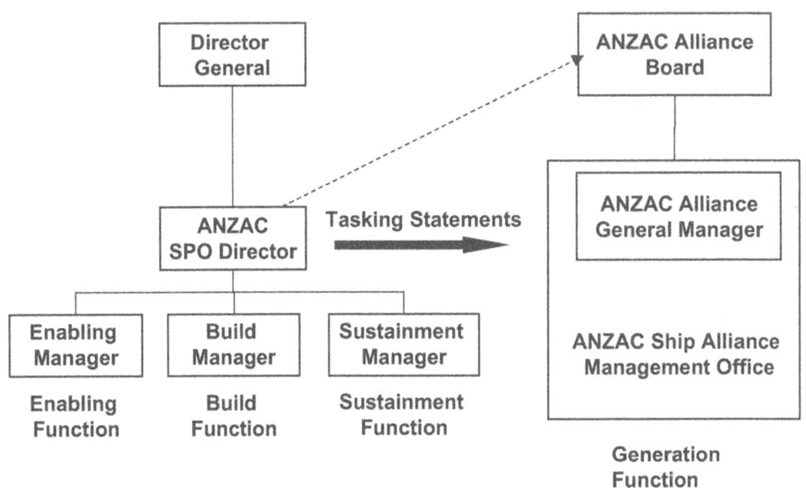

FIGURE 10.4
Tasking process of the ANZAC Ship Alliance virtual enterprise

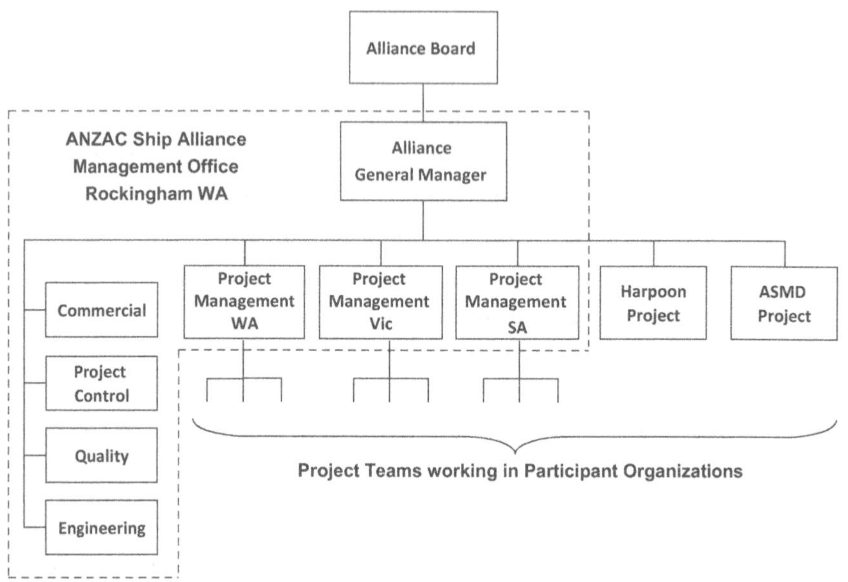

FIGURE 10.5
Organization structure of the ANZAC Ship Alliance

the general manager, commercial managers, and project managers, were seconded from the partner organizations as shown in Figure 10.5. The ASA staff liaised with the partner companies in the spirit of the business model to determine the best outcome for the ANZAC ships.

The project development process followed VERAM. A slightly simplified process was used due to the small number of participants in the ASA (as compared to the first stage of a generic VERAM that is open to any partners). In Figure 10.6, the ASA serves as the coordinating mechanism for the three major partners and its operation is established as the ASA management office (ASAMO).

A task statement triggered the project lifecycle, which is managed by ASAMO staff (i.e., seconded staff dedicated to ASA activities) and drew on appropriate engineering expertise from the partner organizations. This is reflected in the individual project lifecycles of each of the partner organizations.

In this case study, the enterprise was not set up as a legal entity. There was no formal, binding agreement among the partners in the ASA. In the language of virtual enterprises, the partners were loosely linked organizations such that everything done in the ASA was based on trust. The new service functions that were developed on this premise are as listed in Table 10.1.

It is worth noting that at this stage, the ASA can be mapped to elements in 3PE, as shown in Table 10.1. This mapping enables closer examination of the structure and hence the design of the ASA as a system consisting of three organizations (each of which has its own system).

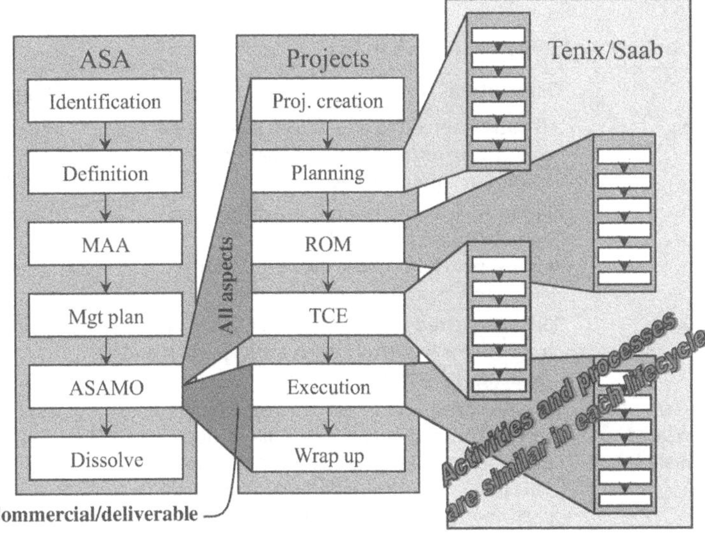

FIGURE 10.6
VERAM of the ANZAC Ship Alliance virtual enterprise

From the point of view of the ship builders, the ASA was an unprecedented business environment in which no one knew exactly how to operate. As outlined previously, when a virtual enterprise is formed there is a need for seamless communication among all partners. The issues with the ASA were primarily related to necessary enhancements of the IT infrastructure to support collaborative work:

- An IT network to provide support to activities both within the ASAMO (Perth) and, where practicable, to support project activities during development and execution anywhere in Australia.
- Access from all ASA work sites, including locations in large cities (such as Sydney, Melbourne, Adelaide) or remote towns.
- In addition to a range of Microsoft Office tools, the deployment of a web-based collaborative toolset on the network.
- The functional specification of a tool to provide document, record, and workflow management within the ASAMO.

In addition, a new financial model was created. Reimbursement to the industry partners was fully open book, subject to verification by audit as a three-limb compensation model:

- Limb 1—100% of what was expend directly on the work, including project-specific overheads.
- Limb 2—A fee to cover all other corporate overheads and profit.

TABLE 10.1

Mapping of Service Elements in ASA to 3PE

Function	Description	Mapped to 3PE
Risk sharing	All parties agreed to win or lose together, be collectively responsible, and share risk and reward in equitable way.	Process
Encourage innovation	Staff in ASA would be outcome driven, make decisions based on "best for project" philosophy, and choose the best resources, skills, and expertise from all parties.	People-Product
Open book	All parties would make sure that all financial transactions were fully open for inspection and challenge by partners in ASA.	Process
Ownership of product-related services rested with ASA	The enterprise was a joint development of several companies and hence the ownership of the product-related services had to be resolved. This issue was eventually settled but as all participants were in the defense business environment, and the customer was a partner of the ASA, the "company" structure on the right-hand side of Figure 10.4 was able to provide sufficient background understanding for the participants.	Process-Product
Secondment of staff from the three partner organizations	Once the "company" status was agreed, it became urgent to develop a set of processes acceptable to all staff seconded from the partner organizations. A lot of time was spent synchronizing practices and cultures from partner organizations. There was confusion in the first year among the staff of the participating organizations about the nature of the ASA. This issue was resolved through a number of ASA workshops.	People

- Limb 3—An equitable sharing between all ASA partners of gain/pain depending on how actual outcomes compared with pre-agreed targets in cost and non-cost performance areas, consistent with the guiding principle that "all parties win or all parties lose."

The idea of this financial model was to compensate the partners according to performance of each project undertaken by ASA:

- Actual costs as recorded in "open book."
- Non-cost items identified and measured with key performance indicators as Limb 3 adjustments.
- Non-cost pool (a pre-agreed amount) set aside for exceptional performance.

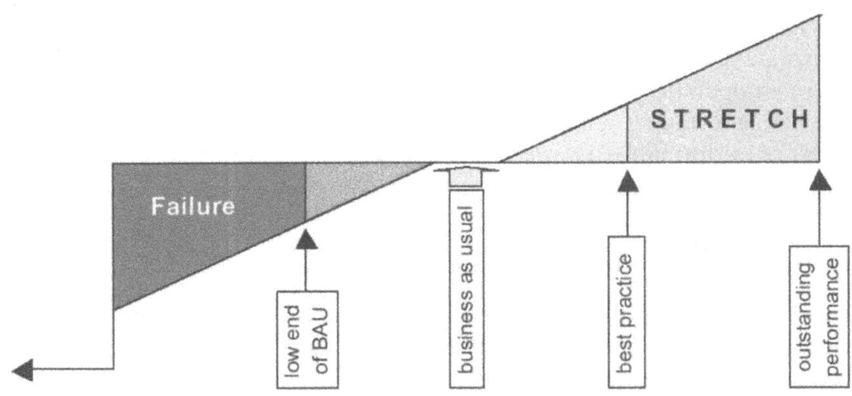

FIGURE 10.7
Performance model of ASA

The rewards were computed according to the business performance model shown in Figure 10.7.

In Figure 10.7, if the project cost fell in the "business as usual" range, the normal (budgeted) project price was awarded. If the non-cost performance was better than expected, the project team was appraised and prizes awarded. If the non-cost performance was worse than expected, the project team was warned.

If the overall performance of the project (including cost and non-cost items) was better than "best practice," the partners were rewarded financially with an agreed amount. On the other hand, if the overall performance of the project was worse than "low end of business as usual," a financial penalty would be imposed on the partners.

10.4 What Are Interacting Networks?

A virtual enterprise consists of many individual enterprises working together. It is a form of system of systems that has a defined goal. Operational networks may be both constrained and liberated by business considerations. Financial resource limitations may condition the nature of network solutions that are viable, and other business considerations may restrict access to some networks.

At the same time, business functions can be viewed from a system-of-systems perspective, as can linkages between businesses that may provide access to scarce resources.

A system-of-systems view draws on business management research studies. Firstly, we take an interaction view (as compared with a resource-based view) of markets, followed by some network concepts framing virtual enterprise collaborations. This is followed by a broader business viewpoint where an operational activity may be linked to financial, technological, and knowledge resources within or external to a particular enterprise. Finally, the matter of performance assessment is discussed, as performance expectations condition system design.

10.4.1 Inter-Enterprise Interactions

Modeling of interactions in systems of systems can be logically represented by an extended formulation of the 3PE model to multiple enterprise networks. The single-enterprise 3PE model has three interaction links within an operating environment: people to process (Pp–Pc), people to product (Pp–Pd), and process to product (Pc–Pd). However, in a multiple-system situation (mostly in organizations), the number of interacting links can expand quickly as given by:

$$l_a = C_2^{3n} \tag{10.1}$$

where n is the number of systems in the system of systems. Moreover, the number of inter-system interactions is given by:

$$l_i = l_a - 3n \tag{10.2}$$

However, if we examine the links carefully, there are in fact only six types of interaction links between systems, as shown in Table 10.2.

The interaction links at the diagonal of Table 10.4 are inter-organizational interactions crossing system boundaries. The nature of these interactions is explained in the following sections.

10.4.2 People-to-People Interaction

General business principles provide a few guidelines governing smooth interactions between enterprises. One of the principles is that people-to-people interactions often occur among peers—individuals with similar organizational status (Figure 10.8).

TABLE 10.2

Inter-Company Interaction Matrix

		System 1		
		People	**Process**	**Product**
System 2	People	(1) Pp – Pp	(2) Pp – Pc	(3) Pp – Pd
	Process	(2) Duplicated	(4) Pc – Pc	(5) Pc – Pd
	Product	(3) Duplicated	(5) Duplicated	(6) Pd – Pd

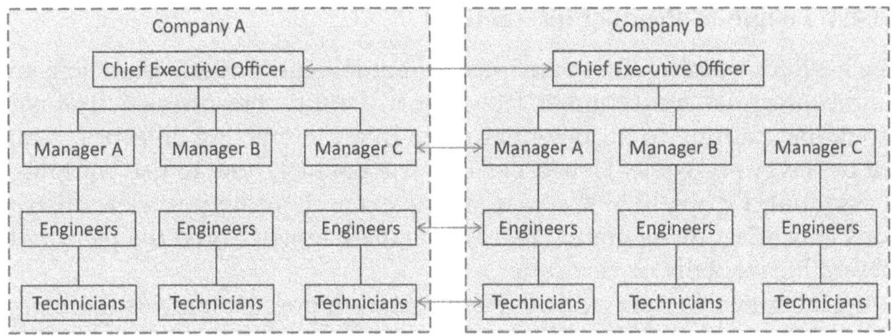

FIGURE 10.8
Peer-to-peer interactions

A reason for this principle is to make sure that communications are received and acted upon by the appropriate level of decision-makers. Hence, the main modeling task for this interaction is about managing the peer-to-peer relationships.

10.4.3 People-to-Process Interactions

The most common process that people across organizations have to deal with is the document process. Different documents have different routes. A document flow model (Figure 10.9) identifies documents and who or where they should go to.

For a system-of-systems scenario, people in one company may not be familiar with the document flow of another company. The document flow model helps people to adapt to processes.

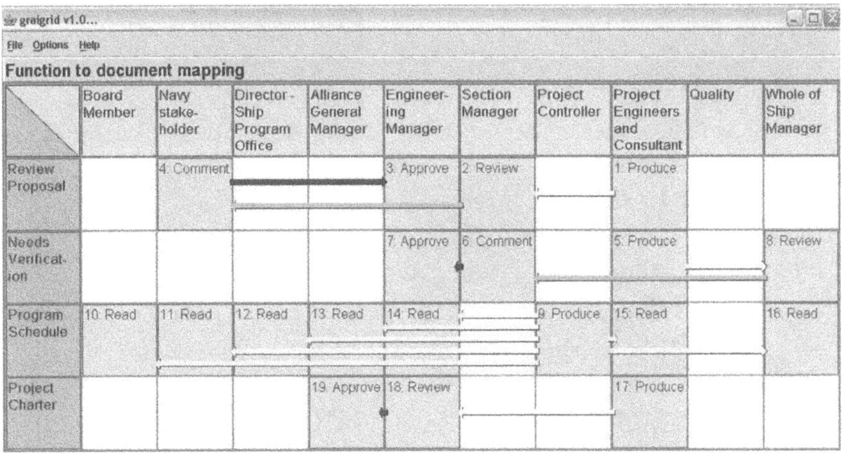

graigrid v1.0...

File Options Help

Function to document mapping

	Board Member	Navy stake-holder	Director - Ship Program Office	Alliance General Manager	Engineer-ing Manager	Section Manager	Project Controller	Project Engineers and Consultant	Quality	Whole of Ship Manager
Review Proposal		4. Comment			3. Approve	2. Review		1. Produce		
Needs Verificat-ion					7. Approve	6. Comment		5. Produce		8. Review
Program Schedule	10. Read	11. Read	12. Read	13. Read	14. Read		9. Produce	15. Read		16. Read
Project Charter				19. Approve	18. Review			17. Produce		

FIGURE 10.9
An example of a document flow model

10.4.4 People-to-Product Interactions

The need for training to use any piece of equipment properly is obvious in a single-enterprise environment. However, in a multi-enterprise environment, providing training to all potential users in all interacting enterprises may not be always possible. Hence, clear instructions on how to use equipment is essential. Figure 10.10 shows a simple example of how unclear instructions may affect the operation of a system of systems. Should the pictures be viewed horizontally or vertically?

Just imagine that a service engineer from a support company is working at a client's location. A fire breaks out and the service engineer has to use a fire extinguisher on the spot. Although occupational health and safety induction is mandatory in many countries, time restrictions can make training in the use of on-site equipment difficult to arrange. Improvements to the instructions, such as an arrow between the pictures to indicate the order would help the user to follow them quickly.

10.4.5 Process-to-Process Interactions

Process-to-process interaction can be modeled by a diagrammatic representation originating from the man-machine chart used by industrial engineers in the Second World War. The modern version is the swimming-lane diagram which

FIGURE 10.10
Instructions for operating a fire extinguisher

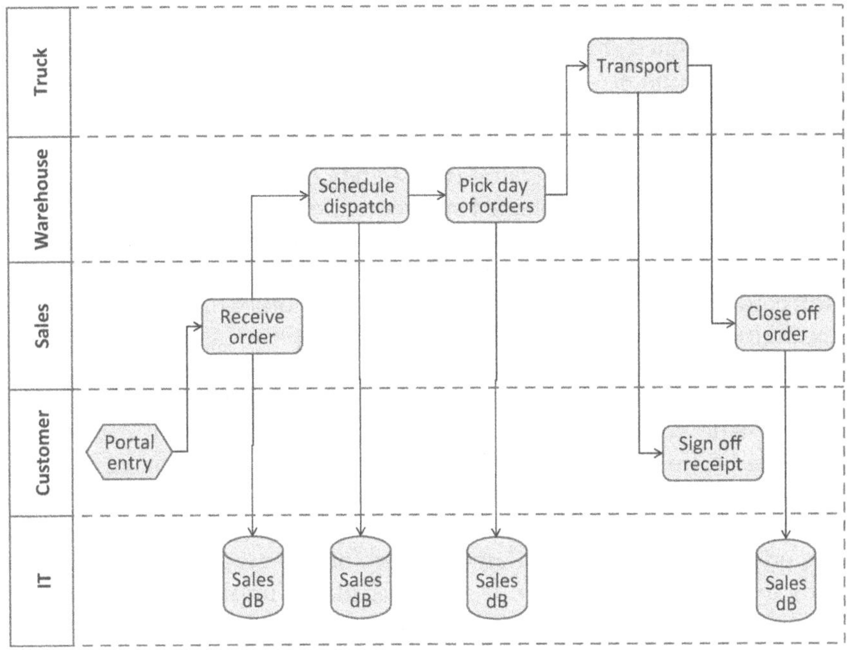

FIGURE 10.11
An example of a swimming-lane diagram

lays out all interacting parties in rows (Figure 10.11). An alternative version is to run vertically in columns. Activities in the swimming-lane diagram are marked and linked so a sequence of activities can be seen between processes.

Process synchronization is not a trivial task. It is often the result of tireless negotiations between the parties. The swimming-lane diagram can show the interactions among multiple enterprises. In that case, a round-table negotiation is necessary.

10.4.6 Process-to-Product Interactions

Although it is possible to change the design of a product to suit special requirements, it is a costly exercise. In process-product interactions, processes can be changed more readily. In any case, a one-off solution could be implemented to fix the problem in the short term while the interacting enterprises are waiting for longer-term solutions to be designed and installed. The term "exception handling" can be used to describe this situation.

An engineering product is designed to satisfy the need of customers. Stakeholder involvement in the decision-making process is absolutely necessary. Fixing a periodic meeting schedule with the customer to ensure all parties are well informed and understood is possible in the product development phase. In addition, synchronized development of the mission system and the support system simultaneously will help the newly changed system to be commissioned smoothly.

Product designs are protected by law. A commonly used definition of design is that design is any composition of lines or colors or any three-dimensional form, provided that such composition or form gives a special appearance to a product of industry or handicraft or can serve as a pattern for such a product. The protected subject matter pursuant to this definition is the individual appearance of the product. A technically functional element is protectable if it also has some aesthetic character and the same technical result could be reached by another shape. On the other hand, design of complex engineering products is substantiated by sophisticated design analysis, which is not always visible.

An exception to design harmonization by features is in mechanical connections. The so-called must-match features are effectively excluded by law related to repair (or spare part). A spare part is a component of a complex product, the purpose of which is to repair the product so as to restore its original features. The necessity to restrict the legal protection of such parts originates from the fundamental idea of liberalizing the movement of goods in the internal market. In practice, this is implemented as a remuneration system for the use of protected designs and the appropriate level for remuneration is stipulated by law. Independent spare-part producers may pay a reasonable remuneration to the rights holder.

Likewise, services for an engineering product can be made exceptional according to the need at the time of service, e.g., when an underlying operating system must be upgraded for a particular reason, and there is a domino effect on other updates that are required at the time. Some vendors will notify customers when their commercial off-the-shelf (COTS) software will no longer be supported, others will not. Notice a key point here—COTS software is dependent on the underlying operating system and hardware, so any COTS software support strategy must be linked to an operating system and hardware support strategy. In terms of the operating system, software drivers and other software applications, e.g., the.Net framework and DirectX, should be considered. There could be consequential impact of changes to other parts of system and hence a holistic support strategy should be adopted. In simple terms, what COTS software relies on and what it affects must be assessed. A support strategy for COTS software cannot be considered in isolation.

10.4.7 Product-to-Product Interactions

Product-to-product interactions occur frequently in service and support operations. For example, several service contractors working together on a system may use their own set of equipment tools to diagnose problems on the system. This means they bring their own "products" to the servicing environment. This situation of equipment brought in from several organizations and used in joint service operations are common.

Support engineers need to be able to generate engineering change proposals (ECPs) that pass the "completeness" test and consider the whole of the system lifecycle. Impact analysis of a proposed change requires consideration of multiple phases of the system lifecycle. So an ECP would need to consider several factors:

- Different system instances at different build states (e.g., all ships of a certain type are not always identical) for significant periods of time.
- Upgrade or disposal of spare-part inventories.
- Staff training for upgraded systems.
- Other factors suggested by the elements of capability.

When developing an ECP, imposing existing current constraints on a new capability should be avoided. The system owner needs to make a judgment call to remove these constraints in order for the new system after ECP to meet operational needs. For example, in the ASA virtual enterprise case, an ECP initiated by defense personnel imposed constraints by mandating the use of in-service communications equipment in a system. Operators of the communications equipment found it difficult and error-prone to set up and maintain the complex communications networks such that eventually the network had to be replaced. It probably would have been more cost-effective to not mandate the in-service communications equipment in the first place. The system owner will often apply their current practices, experiences, and procedures to new capabilities; for example, a naval ship class known as "loading helicopter dock" could be viewed as a transport ship as opposed to being an amphibious fighting vessel, which would then require a different support concept.

Validation of the support requirements used in the acquisition phase (e.g., design for sustainability) is an essential part of an in-service ECP. As the system transitions into service (and as part of monitoring ongoing supportability), the support team should measure support capability, i.e., verify the support system. For example, when introducing a new aircraft into service, an airline should ensure it has some means to perform essential corrective maintenance along the aircraft's intended routes. Before the system is transitioned into service, it is important to objectively measure the capability of support elements, which includes trained staff, spare parts availability, rates of consumption of consumables, etc.

The optimized design solution should include adequate requirements elicitation (requirements collected from multiple viewpoints), complete lifecycle costing modeling, identification and evaluation of alternative solutions, and an assessment of the impact on existing capabilities and infrastructure.

10.5 Case Study

In chapter 4, we used an industry case study of a CNC machine remote operations support system to illustrate the modeling structure of 3PE system [3]. That case is actually modeled as a single system, i.e., the remote operations support system without considering the influence of customers. The environment is taken as an indivisible container, i.e., the universe. This is fine from the point of view of the CNC machine manufacturer and is a sensible approach while concentrating on the technicalities of the system. In this chapter, we extend the model to examine the influence and reactions of customers by separating customers as an independent system which interacts with the CNC manufacturer. In this case, the CNC machine manufacturer is designated the original equipment manufacturer (OEM).

We mentioned in chapter 4 that the CNC machines were configured as servers, with functionality to communicate with the global master server. The IT configuration is represented in Figure 10.12. This functionality allowed the machines to interact with the global master server to provide operations information at client locations.

In this case study, the 3PE inter-enterprise interactions can be identified and modeled as shown in Table 10.3. Using 3PE modeling constructs,

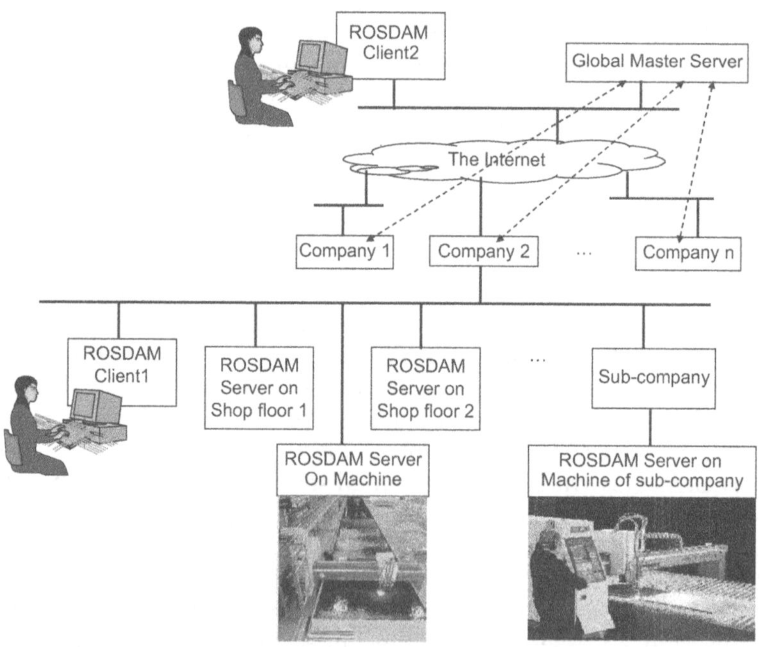

FIGURE 10.12
Remote operations support system showing interaction with customers

TABLE 10.3

Interaction Matrix Between OEM and Customer

		Original Equipment Manufacturer (OEM)		
		People	**Process**	**Product**
Customer	People	Peer to peer: customer relationsKnowledge sharing/transfer; transforming customer data into knowledge. New data-processing algorithms were developed as software modules to perform this function, enhancing operational efficiency.	Adaptation: Engineering information integrated for supporting more effective customer service. Engineering information such as bill of material, machine configuration management, parts inventory, and resources planning were integrated from different sources including CAD, MRP, and various manufacturing sources to create a seamless operations database for the machine.	Training: Communication networks and IT systems based on a client–server model. The normal standalone operating system on the machines was significantly changed to one that could act as a server in a network environment.
	Process	Duplicated	Negotiation: Every machine has some minor differences due to adaptation to the required operational environment. Negotiation between the OEM and the customer is required to ensure the best possible terms acceptable by both parties. The contracting process involving sales and legal services on both sides is an important interaction prior to any further actions taking place.	Exception handling: A customized database for customer's machine. The new system design required the upgrade of field products. Field upgrade for machines that were already installed at customer locations was progressively rolled out according to contracted maintenance schedules.
	Product	Duplicated	Duplicated	Engineering change: New on-machine signal-based diagnostics capability was developed using chaos theory and digital signal processing was developed to assist identification of faults.

the two-company interaction table has nine cells (see Eqn.10.1). Three of these interactions are repeating.

$$l_i = C_2^{3\times 2} - 3n = \frac{6\times 5}{2\times 1} - 3\times 2 = 9 \tag{10.1}$$

Development of new capabilities in the system of systems (i.e., the OEM remote operations support system and the customers) can then be identified according to the interactions. The new support system significantly increased the efficiency of service and support of the machines and reduced operational costs for the OEM.

10.6 Global Operation Support Services

Complex assets are normally built from a large number of components and involve a large number of engineers and contractors [4]. In the past, customers as plant owners usually maintained their own service department. However, the increasing complexity of plant and operating conditions, e.g., environmental considerations, require service personnel to have a higher level of analysis and judgment capability. In managing the design and manufacture of a chemical plant, a service virtual enterprise (SVE) with several partner companies around the world providing after-sales services to a customer was formed (Figure 10.13).

Each SVE partner in Figure 10.13 is an independent entity equipped with its own unique capabilities and competencies, and takes an agreed set of

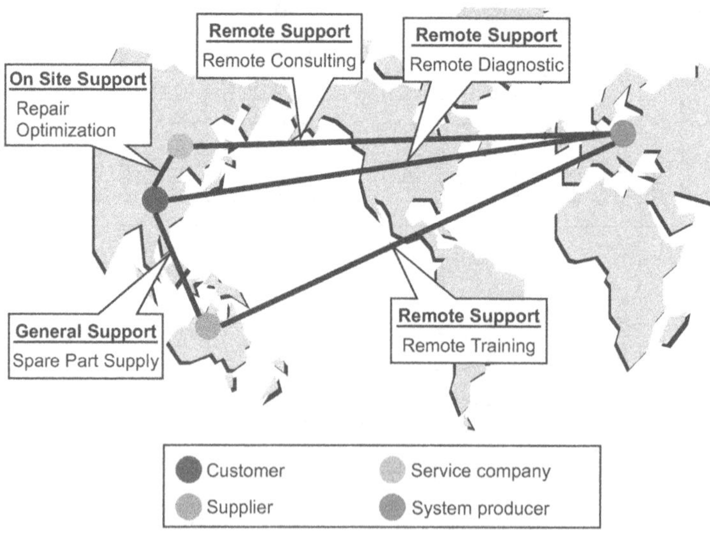

FIGURE 10.13
A global service virtual enterprise

responsibilities to perform the allocated work. The SVE is designed as a "hosting service" that had a broad range of services, including plant monitoring, preventive maintenance, troubleshooting, performance simulation and evaluation, operator training, knowledge management, and risk assessment. Participants in the SVE have well-defined roles and responsibilities.

Essential elements in the design of an SVE are an efficient system architecture and the provision of the right resources to the right service tasks. By synchronizing organizational activities, sharing information, and reciprocating one another's technologies and tools, each partner in the SVE can provide services that would have been impossible individually. This support solution therefore requires properly designed systems to integrate many types of technologies in the provision of support services to customers.

This case has four types of company participating in the network: the OEM, service providers, suppliers, and customers. The number of interaction links is computed in Eqn.10.2.

$$l_i = C_2^{3\times 4} - 3\times 4 = \frac{12\times 11}{2\times 1} - 12 = 54 \tag{10.2}$$

Table 10.4 shows the interaction matrix among all four types of enterprises.

TABLE 10.4

Interaction Matrix Among All Four Types of Enterprises

		OEM			Service Providers			Suppliers			Customers		
		Pp	**Pc**	**Pd**	**Pp**	**Pc**	**Pd**	**Pp**	**Pc**	**Pd**	**Pp**	**Pc**	**Pd**
OEM	Pp				[1]	[2]		[1]	[2]		[1]	[2]	
	Pc					[3]	[4]		[3]	[4]		[3]	[4]
	Pd												
Service Providers	Pp							[1]	[2]		[1]	[2]	
	Pc		D						[3]	[4]		[3]	[4]
	Pd												
Suppliers	Pp										[1]	[2]	
	Pc		D			D						[3]	[4]
	Pd												
Customers	Pp												
	Pc		D			D			D				
	Pd												

[1] The relationship within an SVE differs from a totally authoritative company structure. A much more flexible human organization structure is required to allow collaborations among the partners.

[2] New IT and communication systems were installed to enable inter-company exchange of information as well as personal interaction.

[3] Global access by customers. The SVE was implemented over the internet.

[4] Work items were analyzed individually so that the link from individual level to group level can be streamlined ensuring minimum duplication of work and conflicts.

[D] Duplicated interactions.

10.7 Reflections

Using the 3PE modeling methodology and the network hierarchy, a system analyst can isolate and identify all the relationships in an alliance. Modeling of the relationships depends on the type of relationships involved. Different tools can be used; for example, a swimming-lane model or a sequence diagram in unified modeling language (UML) is normally useful for modeling process-to-process relationships, whereas a factor-rating model is useful for people-to-people relationships. At this stage, the most important action is to determine these relationships and define their boundaries.

You are now encouraged to use the 3PE modeling methodology to isolate and identify an alliance from your experience. The alliance, by definition, must have at least two companies (or systems) that can be individually modeled by the 3PE model. Working in a collaborative team can be a tremendously challenging situation. The project leader attracts a lot of attention from different directions. This exercise gives you an opportunity to reflect on your experience in a team environment and identify good and bad situations. To begin, develop an interaction matrix similar to Figure 10.6 and populate it with your findings.

10.8 References

1. Mo J.P.T., Nemes L. (Eds.) (2001). Global Engineering, Manufacturing and Enterprise Networks, pub. Kluwer Academic Press, ISBN 0-7923-7358-8.
2. Mo J.P.T., Zhou M., Anticev J., Nemes L., Jones M., & Hall W. (2006). "A study on the logistics and performance of a real 'virtual enterprise'." *International Journal of Business Performance Management*, 8(2–3), 152–169.
3. Mo, J.P.T. (2003). "Case Study - Farley Remote Operations Support System." In *Enterprise Integration Handbook*, Bernus, P., Nemes, L., & Schmidt, G. (Eds.) Springer-Verlag, Chapter 21, 739–756. ISBN: 3-540-00343-6
4. Kamio, Y., Kasai, F., Kimura, T., Fukuda, Y., Hartel, I., & Zhou, M. (2002). "Providing Remote Plant Maintenance Support through a Service Virtual Enterprise." In *VTT Symposium 224, Global Engineering and Manufacturing in Enterprise Networks, 9-10 December, Helsinki, Finland*, pp.195–206.

10.9 Additional Reading

Mo J.P.T. & Nemes L. (Eds.) *Global Engineering, Manufacturing and Enterprise Networks*, Kluwer Academic Press, 2001. ISBN: 0-7923-7358-8

11

Markets and Service Ecosystems

The blockchain concept was pioneered within the context of cryptocurrency Bitcoin, but engineers have imagined many other ways for distributed ledger technology to streamline the world. Stock exchanges and big banks, for example, are looking at blockchain-type systems as trading settlement platforms.

Anthony Scaramucci
American financier, 2016

11.1 Some Matters of Context

We consider a marketplace (as compared with markets) as an environment where different things may be bought and sold, with some areas or regions specializing in a particular product or service, and where buyers and sellers may compete or cooperate. We can buy components to build things ourselves, or we can buy something already assembled. Or we can just browse and appreciate what is available. We may do this in a physical marketplace or online.

Viewed as a mission, our shopping may be associated with several actions:

- Trading. There may be a change of ownership following a transaction, or access to goods or services may be provided for a fee, e.g., renting a car. A variety of protocols are associated with trading—free or restricted access, accepting or rejecting a fixed price, negotiating a price that may include bundling of goods and services, competing in an auction, or simply registering a best and final offer. Some of these human practices are emulated on eBay, for example, where background IT systems support the process.
- Exploring/mapping. Individuals and organizations may prepare for future trading by collecting and organizing what is learned about the possibilities and trends, or searching for the best option on something we are interested in. The authors see the emergence of

TABLE 11.1

Some Market-Oriented Activities

Environment or Trading Orientation	Mission or Market Activity	
	Trading	Exploring/Mapping
Financial assets	Stock market shares	Following trends in share prices
Physical assets	Consumer goods	Comparing features, finding availability
Intangible assets	Entertainment	Making catalogs, searching for tickets
Intellectual assets	Education	Establishing a knowledge base

Big Data technologies to collect, store, analyze, and manipulate huge volumes of data from diverse sources, mapping and making predictions about things that are not yet clearly visible. People use on-line tools like Google to search for all kinds of things on a daily basis. Complex background systems of systems make this easy for us.

The environment attributes of a particular trading event are usually related to the type of assets being traded:

- Financial assets being traded through stock exchanges and banks where financial activity mapping and price/cost trends are important for both buyers and sellers: Where can I get access to funds and on what terms? What are the trends in interest rates or share prices?
- Physical assets being traded directly by manufacturers or through retail outlets, being offered for rent or being purchased with the aid of a broker, e.g., in real estate: What is available, where and under what conditions?
- Intangible assets like patents, software, entertainment events, or personal health: What is available, where and when?
- Intellectual assets, e.g., knowledge held by people working as contractors or being shared through educational institutions or collaborative ventures.

Mapping these environment attributes against mission components (Table 11.1) reveals a variety of scenarios where instances of systems of systems may be observed.

11.2 A Stakeholder Perspective

It has been observed that as a particular market for a product, service, or new technology evolves, there is a typical maturity cycle with different groups of

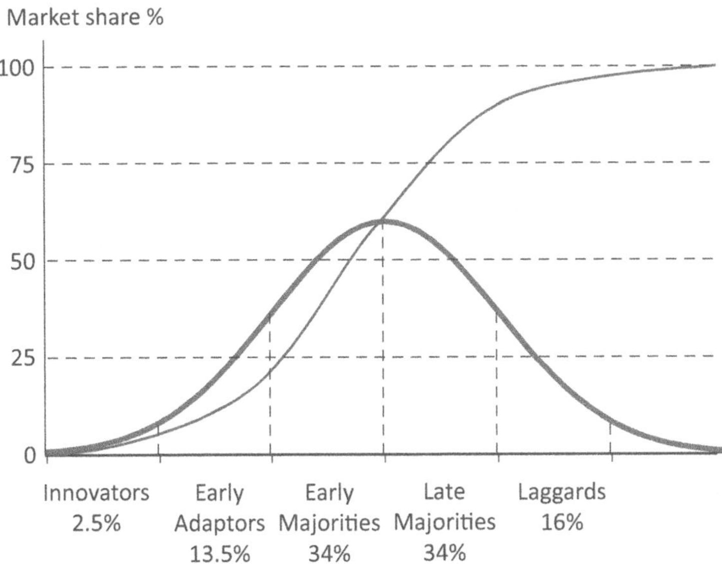

FIGURE 11.1
Market maturity cycle

stakeholders becoming dominant at each stage (Figure 11.1). Each group has its own characteristics and dominant logic. Innovators and early adopters may look for lots of features they can experiment with, and may be interested in interacting with the technology developers. The early majorities are only interested in particular features that have some utility for them, and the late majorities will focus on features that have be shown to work reliably. During the majority market phase, more competitors may emerge, offering different optimizations of features and competitive pricing. The laggards are most commonly interested in price.

This can mean that through the lifecycle of a product or system, the focus of innovation may change from relative novelty to reliability to cost of production and operation. One of the authors was advised by a marketing executive that "the resolution of last year's problem is next year's selling feature." These considerations fall into the category of "concerns" discussed earlier in chapter 3. This pattern needs to be understood in the design and operation of systems of systems and is reflected in working with lead users and releasing beta versions to get started. But there needs to be an understanding that while working with these stakeholders is necessary, it is not sufficient, for they account for less than 20% of the market. That is unless the asset or service is only relevant to the early group, e.g., the provision of system development tools, and even in this niche market, the same kind of pattern will be observed.

There are other kinds of generic actors:

- Competitors. The systems developed by competitors can set benchmarks for utility, cost and performance expected by users, and study of their offerings can identify gaps or where better functionality can be delivered.
- Complements. Systems developed by others that may complement yours, illustrated in the interplay between smartphone providers and the developers of apps.
- Governments and institutions like standards bodies that may set rules for system testing and operation.
- The broader community, whose values may influence the market for particular products and systems.

Markets are associated with generic classes of concerns:

- Price—initial cost and the cost of access and ownership.
- Delivery—speed and location of initial delivery and of support services.
- Quality—this has many dimensions, including fitness for purpose, durability, reliability, ease of maintenance, ease of upgrading, how the product is valued, and associated expectations (e.g., environmental sustainability).

11.3 Viewpoints and Models

It is observed in the ISO 42010 standard that to properly understand the operation of a complex system of systems, it needs to be viewed from multiple perspectives. A viewpoint is defined in the standard as a "specification of the conventions for constructing and using a view. A pattern or template from which to develop individual views by establishing the purposes and audience for a view and the techniques for its creation and analysis." In the following sections we present ideas about patterns or templates that have different origins, but all combining market and system thinking.

11.3.1 Markets as Networks

A group of European researchers, the IMP Group (http://www.impgroup.org), began to explore the utility of viewing markets from a network perspective in the 1980s. They framed markets as intersecting networks of activities, actors, and resources linked together in particular ways (by resource ties and

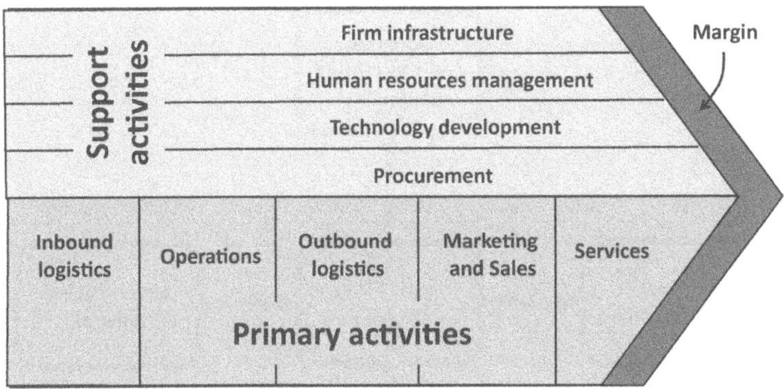

FIGURE 11.2
Enterprise value chain

activity links) for particular "missions." In using this concept, the authors have observed that while there may seem to be dynamic change in the marketplace, this is associated more with reconfiguration of linkages within and between the three networks rather than rapid changes in the networks themselves. This means that mapping each network associated with a particular mission can provide a good as-is foundation from which to explore to-be situations in response to different stimuli, such as the introduction of a new transport infrastructure.

The activity network may be viewed as interlinked functional business activities, reflecting a version of Michael Porter's enterprise value chain, illustrated in Figure 11.2. Or the activities may be those associated with a particular element of such as a particular supply chain supporting the provision of inputs to a particular enterprise, or the macro-level supply chain associated with a particular industry, e.g., the agriculture or automotive sector. This network indicates what has to be done, without necessary specifying how it will be done.

Mapping an actor network starts by considering generic actors—institutional actors like governments, platform providers like telcos, and enterprise actors represented the five market forces illustrated in Figure 11.3 [1]. While the actor network may be relatively stable, it can be disrupted by the introduction of a new actor, alternative ways of servicing a function, or by a change in institutional actor rules that may also introduce a new activity such as environmental impact monitoring. Strategic initiatives may change linkages, e.g., traditional competitors or enterprises and institutions collaborating for some purpose.

Mapping a resource network entails identifying assets associated with each enterprise function, external infrastructure that may be drawn upon, and their interconnections. It may be appropriate to consider both tangible and intangible assets (e.g., domain knowledge networks). In our experience, resource element networks change relatively slowly, as it takes time

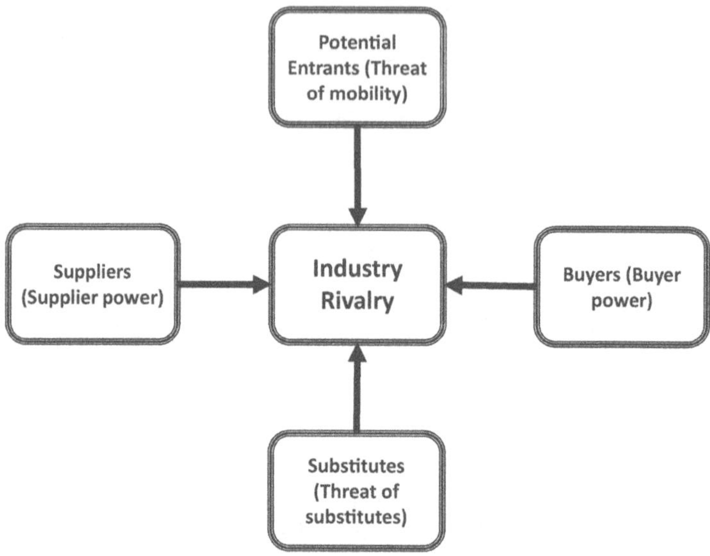

FIGURE 11.3
Five network forces

to acquire and bed down a new asset or infrastructure. But there comes a time where a critical mass may develop that changes how things work, as evidenced by the explosive growth in the adoption of GPS technology as the global coverage and accuracy of the technology evolved. Resource elements may provide a foundation for innovation, as can the creation of novel linkages between them as the capacity and capability of each resource evolves over time, e.g., in the way IT and transport infrastructure interact.

The three networks are linked to varying extents by information flows and physical flows, depending on the extent to which the mission is the supply of goods or services where matters of volume and variety need to be taken into account.

It is our experience from mapping complex systems that the thinking represented here can be useful as a precursor to the identification and development of a system architecture description. Simply using structured questions about the nature of the three networks and related flows provides a good contextual foundation for subsequent design and operational considerations.

11.3.2 Markets as Complex Adaptive Systems

In viewing markets as complex adaptive systems, autonomous agents are assumed to choose what to offer for sale, what to buy, where and when to buy or sell and how to facilitate or inhibit specific kinds of transactions. As well as the system (the market) adapting to changing conditions, the agents can also adapt through a process of learning.

Koritarov [2] described the construction and operation of a simulation of an electricity generation and distribution system using agent-based modeling and simulation, with different kinds of agents having their own decision rules interacting within and between different system layers. Different planning periods were represented depending on the level of detail sought. A special-event generator was used to simulate power outages. This tool provided the ability to investigate complex interactions between the performance of the physical generation, transmission, and distribution infrastructures and the economic behavior of market participants under different conditions.

Rouse [3] studied the US healthcare system, starting with R&D and regulation, through pharmaceutical and equipment providers and hospitals, then the community, forming the view that decomposing and recomposing elements of the system was not an appropriate approach to seeking improvement in a complex dynamic environment composed of largely independent agents. He viewed this market as a complex adaptive system with the following attributes:

- They are nonlinear and dynamic and do not inherently reach fixed-equilibrium points.
- They are composed of independent agents whose behavior is based on physical, psychological, or social rules rather than the demands of system dynamics.
- Because the needs or desires of the agents, reflected in their rules, are not homogeneous, their goals and behaviors are likely to conflict. In response to these conflicts or competitions, agents tend to adapt to each other's behavior.
- Agents are intelligent. As they experiment and gain experience, agents learn, and change their behaviors accordingly.
- Adaptation and learning tend to result in self-organization. Behavior patterns emerge rather than being designed into the system.
- There is no single point(s) of control. System behaviors are often unpredictable and uncontrollable, and no one is "in charge." Consequently, the behaviors of complex adaptive systems can usually be more easily influenced than controlled.

Reflecting on these attributes in studying the attributes of several complex market sectors, including healthcare [3], we make the following observations. Firstly, the nature and extent of the value placed on business-to-consumer transactions influences the whole value chain. If there is no perception of value, there is no reward for any of the contributing actors. Secondly, the ratio of complexity at the consumer end to the total system complexity reduces with market maturity and the initiatives of successful players. In other words, complexity is increased where it can be managed best.

From the experience gained in analyzing complex systems this way, Rouse [3] suggested that traditional management command and control would be less effective than leadership providing balanced incentives and inhibitions with a focus on outcomes, agility, and personal commitments. He suggested information to oversee the system should include measurements and projections of system state and performance, observations about system stakeholder engagement and performance, and capabilities for exploring "what if" questions.

Paina and Peters [4] used a complex adaptive systems lens to explore the impact of different assumptions influencing evolutionary pathways to achieve an increased scale of health system delivery in developing countries. This followed the observation that blueprint approaches did not fit the dynamic ways in which such health services evolved. One lesson learned from their modeling was the need to pay more attention to local context, incentives, and institutions. The phenomena they focused on were path dependencies, feedback, scale-free networks, emergent behavior, and phase transitions, and how they might be observed in the health sector.

11.3.3 Exploring Market Architectures

In their study of multirobot coordination, Dias et al [5] make some observations:

> Given a team of robots, a limited amount of resources, and a team task, researchers must develop a method of distributing the resources among the team so the task is accomplished well, even as teammates' interactions, the environment, and the mission change. Humans have dealt with similar problems for thousands of years with increasingly sophisticated market economies in which the individual pursuit of profit leads to the redistribution of resources and an efficient production of output. The principles of a market economy can be applied to multirobot coordination.

Such studies are treated as market simulations of a virtual economy where the robots are traders, tasks and resources are traded commodities, and virtual money acts as currency. Robots compete to win tasks and resources by participating in auctions that produce efficient distributions based on specified preferences. When the system is appropriately designed, each robot acts to maximize its individual profit and simultaneously improves the efficiency of the team. There may be specialist negotiator agents described as "traderbots." As with markets, success may be associated with the ability to respond to dynamic changes.

This view of markets draws attention to the kinds of trading processes that are used in markets: dyadic negotiations, triadic negotiations involving a broker or moderator, auctions, best-and-final responses to time-limited offers, set price take-it-or-leave-it offers, and centralized vs. decentralized decision-making.

Emulating these processes using artificial intelligence (AI) techniques can result in the evolution of intelligent agents for a variety of circumstances, e.g., in automating some stock exchange transactions.

11.3.4 Service-Dominant Logic

As noted earlier, markets are associated with trading, involving a buyer-seller exchange of some sort. In the service-dominant logic viewpoint, it is argued there is some form of negotiation where a service entity and a client entity co-create value through the process. The client entity realizes value through use of the tangible or intangible asset accessed. The service entity realizes value through the financial transaction, or in the case of an organization providing community services, from the achievement of a community goal.

In applying service-dominant logic to healthcare operations, Lillrank et al [6] framed value co-creation as a micro-level service event that links healthcare operations management to service-oriented architectures and service-dominant logic. They describe a series of linked events as a patient health episode, where there were interactions with higher-level aspects of the health management system (e.g., managing the availability of facilities). They observed that "process management is appropriate in situations where there is a structured flow with a sufficient volume of similar repetitions. In the case where there are significant amounts of exceptions, a process can be decomposed into service events that can be defined and managed as part of a supply chain". The point here is that one service event (e.g., initial assessment of a patient) may lead to a variety of subsequent events (e.g., running tests or providing medication).

The primary proponents of service-dominant logic outlined some foundational premises, which have evolved over time. A 2016 representation is shown in Table 11.2.

In discussing the addition of premise FP11, Vargo and Lusch [7] observed that while the core idea is associated with individual and dyadic structures (e.g., in B2B or B2C transactions) at a micro level, it can be applied at higher levels of analysis. At a meso level (e.g., seeing a patient episode as an event in the health system), resource integrators are involved, and at a macro level, clusters of resource integrators, institutions, and associated rule structures may be viewed as a services ecosystem. They also suggest that an activity at one level can only be understood by also viewing it from other levels.

Drawing on our own experience of using service-dominant logic concepts, we have framed the process view as shown in Figure 11.4.

TABLE 11.2

Foundational Premises of Service-Dominant Logic (adapted from Vargo and Lusch [7])

	Premise	Explanation/Justification
FP1	Service is the fundamental basis of exchange.	It revolves around processes such as a sales negotiation process. Exchanges may provide access to tangible, intangible, or service resources.
FP2	Indirect exchange masks the fundamental basis of exchange.	Goods, money, and institutions mask the service-for-service nature of exchange.
FP3	Goods are distribution mechanisms for service provision.	Goods (both durable and non-durable) derive their value through use—the service they provide.
FP4	Operant resources are the fundamental source of competitive advantage.	The comparative ability to cause desired change drives competition.
FP5	All economies are service economies.	Service (singular) is only now becoming more apparent with increased specialization and outsourcing.
FP6	Value is co-created by multiple actors, always including the beneficiary.	There are direct and indirect interactions between actors in negotiating and delivering what is valued.
FP7	Actors cannot deliver value but can participate in the creation and offering of value propositions.	The firm can offer its applied resources and collaboratively (interactively) create value following acceptance, but cannot create or deliver value alone.
FP8	A service-centered view is inherently customer oriented and relational.	Service is customer-determined and co-created; thus, it is inherently customer oriented and relational.
FP9	All economic and social actors are resource integrators.	Resources include operant resources (e.g., knowledge, skills) and operand resources (e.g., materials and infrastructure). Integration implies drawing combinations of these things together.
FP10	Value is always uniquely and phenomenologically determined by the beneficiary.	Value is idiosyncratic, experiential, contextual, and laden with meaning. Different individuals or groups may value different aspects of a service more highly.
FP11	Value co-creation is coordinated through actor-generated institutions and institutional arrangements.	Institutions can include such things as property rights, norms, and monetary systems.

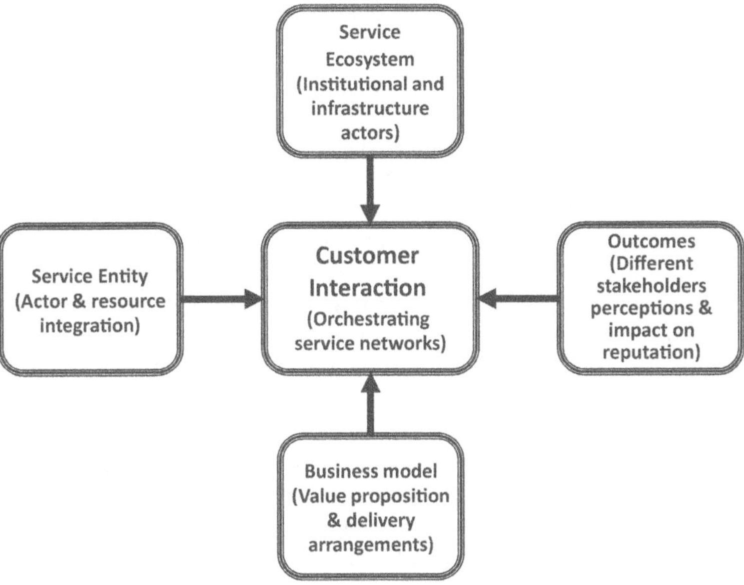

FIGURE 11.4
Five-network focus

11.4 Case Study

Our case study is a Singapore-based bank, DBS Bank, which we will view through the ISO 42010 lens outlined in the preceding sections of this chapter, drawing on publicly available information. We draw on statements by some of the Bank's executive management to help draw out the rationale for the system architecture they chose. The Bank had been in a situation where it had to change its market position, or become uncompetitive. DBS Bank has now received worldwide recognition for its digital agenda, becoming the first bank to be named the world's best digital bank at the prestigious Euromoney 2016 Awards for Excellence.

> At DBS, we believe that banks tomorrow will look fundamentally different from banks today. That's why we have spent the past three years deeply immersed in the digital agenda. This has been an all-encompassing journey, whether it is changing the culture and mindsets of our people, re-architecting our technology infrastructure, or leveraging Big Data, biometrics and AI to make banking simple and seamless for customers.
>
> **Piyush Gupta,**
> *CEO of DBS Bank, 2016*

DBS has strong market positions in consumer banking, treasury and markets, asset management, securities brokerage, and equity and debt fund-raising in Singapore and Hong Kong. From 2006, DBS began to introduce particular data security mechanisms in support of internet banking. The bank's strong capital position, as well as credit ratings that are among the highest in the Asia-Pacific region, earned it Global Finance's "Safest Bank in Asia" accolade for six consecutive years, from 2009 to 2015.

In relation to mission, banks may offer both trading and browsing services—being able to make deposits and withdrawals by various means, being able to check the status of accounts, or explore the repayment arrangements for a home loan online. To kick off its transformation, DBS chief executive officer Piyush Gupta instituted a new corporate strategy centered on the concept of "Asian-ness." He said, "In the next five years, our industry is going to go through cataclysmic disruption. Within the next decade, there will be banks that will make the transition and banks that will die." DBS had to transform itself to be as agile, innovative and fast-to-market as the famed "GAFA" companies of Google, Amazon, Facebook and Apple (and in Asia, you can add Alibaba). While simultaneously living with the reality of being a tightly regulated industry, DBS also competes with hundreds of innovative fintech start-ups that pick off some of the most profitable areas of banking.

Adopting the digital bank strategy, DBS have been able to operate effectively in a non-traditional environment, drawing on emergent technologies to work with very-low-income clients that might not normally have a bank account; accessing the so-called bottom-of-the-pyramid market where huge numbers of transactions for very small amounts are the norm.

Concerns from a banking perspective related to the means to access such a market, the reliability of the technology adopted, and the economics of engaging with very-low-income clients. The stakeholders in the DBS case were the clients in this market sector, existing DBS clients and staff, the DBS shareholders, partner technology providers, and governments.

While the bank operates in many parts of Asia, in this case study we will focus on its entry into the Indian market. DBS established Digibank in India as a global first because of the significant digital infrastructure in India. The whole India infrastructure—the JAM (Jan-Dhan, Aadhaar, and mobile) trinity—and the India stack is under-appreciated. The India stack refers to four government initiatives—biometric authentication, digital records, cashless transactions, and digital consents where acceptance can be acknowledged without a "wet" signature.

The bank established a presence in India without opening any branch offices. A DBS Digibank account can be opened by downloading an app and providing a fingerprint authentication at any one of 500 designated Café Coffee Day outlets across the country. The bank is in talks to have more centers for biometric authentication. According to Gupta, DBS will always be in a position to offer a better deal than others in minimum balance requirement and on interest rate as its costs are lower. Not only does the bank avoid branch and people costs, it is spending a fourth of what conventional banks spend on back-office operations. One reason for this is that it is using robots that act as virtual assistants. The bank is the first to deploy experimental conversational technology from Kasisto, a spin-off from the Stanford Research Institute that created Siri. "There is recognition that cognitive technology is becoming real much faster than people have thought. Field after field, computers are beginning to mimic human beings fairly well," said Gupta. He also suggested that if you think about technology developments, firms like Amazon.com and Netflix have stimulated the development of cloud-based architecture, which also helps re-imagine how business works.

Considering this case from different viewpoints, we start with the markets-as-networks viewpoint (11.3.1). The range of functional activities undertaken by DBS did not change, but the actor and resource networks and some connection between them did. The needs of the low-income client base were different from those of traditional clients.

There were new or different connections within the actor network. While moving toward branchless banking might result in staffing cuts, the move allowed staff related to be re-assigned to expand higher value services offered by the bank like asset management. One example given was the cross-border remit service where traffic almost doubled over a two-year period, increasing revenue whilst reducing cost. Training included involvement in a 72-hour "hackathon." A DBS HR director observed, "So what started as a way of 'training' to be more digital has ended up as a very powerful product ideation process and, in fact, two of the prototypes have already been built into full apps and are about to be launched, with another eight under review by sponsors." DBS ran a number of five-day programs, each beginning with a digital master class, and then leveraging human-centered design and "lean start-up" thinking in a hackathon with real start-ups. Talent from one of the winning teams described the experience as "mind-changing." One said he had never realized he had been seeing the world in analogue until he "saw it all in digital." Like black-and-white TV, once you've seen color TV, you can never go back. So this formula of making senior bankers act like a start-up for 72 hours has had a real long-term impact and helped them pass the tipping point of "getting digital." The HR Director also observed that young talent, who had been through the boot camps and hackathons were empowered to reverse-mentor senior leaders.

The DBS Chief Technology Officer suggested that using a digital distribution approach, DBS could launch into countries with large population—India, Indonesia, China—in a way that it could not through a brick-and-mortar approach. He said "The brick-and-mortar strategy requires many years of payback and very deep pockets – 10, 20, 30 years. But a digital strategy could allow you to access the retail market at a substantially better cost point, and get scale in a way that is actually affordable."

The Indian government was a significant actor, addressing its own background concerns to provide an environment and infrastructure that helped

DBS enter the Indian market. We comment on two such concerns here. Firstly, the government mechanisms for providing financial assistance to low-income individuals were through intermediaries, as the intended beneficiaries did not have bank accounts. But this system was open to fraud. Secondly, there was a substantial cash economy operating that reduced the amount of tax collected. The government took the dramatic step of cancelling the banknote most commonly used in this cash economy, with the only way to recover the funds being a credit to a personal bank account. This promoted the need for a clear means of individual identification and the opening of a bank account.

New resources drawn upon in India were a biometric-based ID card, the expanding Indian ICT infrastructure and talent pool, and chains of convenience food outlets. Combined with the DBS cloud and app-based infrastructure, this meant that a traditional activity, opening a bank account, could be handled differently. In India the bank is offering digital bank accounts with zero balance requirements, 7% interest rate on savings, and unlimited free access to ATMs. These accounts can be opened by anyone with a smartphone using a biosecurity identification tool and a national ID card. A variant of the offering has been established to help small companies get started, and an app called "business class" offers access to financial tools and market insights.

Introducing the viewpoint of markets as **complex adaptive systems** (11.3.2), the previous case discussion illustrates how the autonomous agent behavior (that of the Indian government) can change the rules, and how the addition of a new actor in the marketplace (low income clients) can change the nature of the opportunities. Taking this viewpoint, humans, software, and physical things can act as agents. The DBS chief technology officer (CTO) observed:

> Most banks have large systems which date back 40 or 50 years ago. Maybe even longer than that. Then if you try to offer something to the customer from these systems, you've got to write the software code, every time from the backend system to some frontend application. And each time you want to make a change, you have to re-write the code. So this makes the whole process very clunky. At DBS, we created a middle layer called an Application Programming Interface (API) layer. So now our backend systems only publish to the API layer (a new agent), and we use it to connect to the front. And the chain, instead of going all the way from the back to the front but via a middle layer, makes us a lot more nimble and speedy.

The CTO defied conventional wisdom: "I told our innovation team: don't innovate," he says. "Instead, teach the rest of the organization to innovate." The strategies adopted resonate with those proposed in relation to the health system in section 11.3.2—make things simpler for the consumer by increasing support complexity in a manageable way. The CTO observed that people's lives do not revolve around banking: "If you're making a major purchase like buying a refrigerator, the smaller and faster the banking piece, the better. To make banking joyful, make the banking part invisible."

Introducing the market-architecture viewpoint (11.3.3), we note that DBS leverages AI-driven automation in order to employ a tiny fraction of the back-office staff that a conventional bank normally requires. The bank builds on its wide use of experiences based on market ecosystem "plays"—negotiated transactions organized in different ways for different purposes. The CTO commented: "This digital ecosystem of partners and other third-party participants in the customer journey is an essential part of DBS's digital story. As banking becomes commoditized, we need to control the ecosystem."

From the service-dominant logic viewpoint (11.3.4), it can be seen by referring to Figure 11.4 that both the service ecosystem and the value proposition influence the negotiated service accessed. The service entity is an intelligent agent, and the service ecosystem is a combination of the bank IT infrastructure, local internet access capability, and physical infrastructure (e.g., ATMs). The DBS value proposition not only offers a low cost on-line banking service, but an unconventional means of opening a bank account and secure access arrangements. The DBS website lists service options customers can access:

- Check your balance with a swipe. No login required.
- Your fingerprint is your password. Just touch to unlock your account.
- Add up to eight Quicklinks to directly access your most frequently used services.
- Receive personalized research and insights at your fingertips.
- Here's how we keep you safe while banking with us: 100% confidence with automatic coverage under our Money Safe Guarantee—in the unlikely event that there is an unauthorized transaction, we will repay the money taken from your account.
- Safe and secure transactions with Transaction Signing and (multi-factor authentication) & transaction alerts.

As noted earlier, "going digital" has offered benefits to multiple stakeholders—customers, shareholders, and the government. In the service-dominant logic viewpoint, maintaining a good reputation is seen as important. The CTO observed that six months were spent narrowing down the definition of "Asian service" to respectful, easy to deal with, and dependable. This led to the acronym "RED," an adjective that became part of the corporate vocabulary. To help the bank become RED, the CTO instituted a series of process improvement events: five-day cross-functional workshops focused on eliminating waste—in terms of wasted customer time. The benefit to customers was seen as dramatic. "One year later, we had the top customer satisfaction scores in Singapore." And as noted earlier, DBS use a variety of initiatives to build and maintain trust.

11.5 Reflections

In this chapter, we have regarded market behavior like that of a complex adaptive system, and viewed contributions from social science and management researchers through an architectural-description lens (see chapter 3). Thus, popular ideas become viewpoints, and we have discussed our case study from multiple viewpoints. We have noted that markets are places where assets can be traded to facilitate value creation. We have also noted an increasing interest in automated trading using intelligent agents that emulate the practices of experienced human negotiators. Our case study provides insights into the transformation of one organization that has comprehensively embraced digital technology. Drawing on your personal experience, identify a market-driven enterprise, and think about these questions:

- How would you describe the mission of the enterprise that reflects the kinds of market transaction it is involved with?
- How would you describe its operating environment and any associated concerns?
- What kinds of viewpoints would help understand the operations and competitive position of the enterprise?
- What opportunities do you see for this enterprise to adopt emerging digital technology tools?

11.6 References

1. Porter, M.E. (2008). "How competitive forces shape strategy." *Harvard Business Review*, January, pp. 25–41.
2. Koritarov, V.S. (2004). "Real-world market representation with agents." *IEEE Power and Energy Magazine*, 2(4), 39–46.
3. Rouse, W.B. (2008). "Health care as a complex adaptive system: implications for design and management." Washington, DC: National Academy of Engineering, *The Bridge*, 38(1), 17.
4. Paina, L. & Peters, D.H. (2011). "Understanding pathways for scaling up health services through the lens of complex adaptive systems." *Health Policy and Planning*, 27(5), 365–373.
5. Dias, M.B., Zlot, R., Kalra, N., & Stentz, A. (2006). "Market-based multirobot coordination: A survey and analysis." *Proceedings of the IEEE*, 94(7), 1257–1270.
6. Lillrank, P., Groop, J., & Venesmaa, J. (2011). Processes, episodes and events in health service supply chains. *Supply Chain Management: An International Journal*, 16(3), 194–201.

7. Vargo, S.L. & Lusch, R.F. (2016). "Institutions and axioms: an extension and update of service-dominant logic." *Journal of the Academy of Marketing Science,* 44(1), 5–23.

11.7 Additional Reading

Easley, D. & Kleinberg, J. *Networks, Crowds, and Markets: Reasoning about a Highly Connected World,* Cambridge University Press, 2010.

Swan, M. *Blockchain: Blueprint for a New Economy.* O'Reilly Media, 2015.

12

Business Models Making Operational Sense

> A business model articulates the logic and provides data and other evidence that demonstrates how a business creates and delivers value to customers. It also outlines the architecture of revenues, costs, and profits associated with the business enterprise delivering that value.
>
> **David Teece, 2010**

12.1 Perceptions of Value

If our sole focus was on economics and functionality, we would all be buying one of a few models of automobile. There would be no luxury cars or high-performance cars or electric/hybrid vehicles. The point here is that there are different perceptions of value related to social, economic, and environmental considerations. Doing business is about negotiating a deal where the seller is able to extract some value, and the buyer believes the deal offers better value than the alternatives at the time of purchase, but also considering the later value in use that may be realized. For example, a buyer may be prepared to pay a higher price for a longer-lasting or more reliable product. The same considerations apply to the provision of services. Outright sale or rental of a product may be offered, or products and services may be bundled together, e.g., in the provision of a fully maintained lease vehicle.

An enterprise creates and delivers value through the combined activities of many functional specialists that in some way add value directly or through supporting others in constructing a product or system that transforms or combines the inputs of external suppliers. So, viewed through a value lens, any product, system, or service provider has to engage with particular users and understand how they realize value from its use, and also establish a capability to deliver the product, system, or service in a sustainable way. Value structures include value creation and value delivery considerations.

How all this works together is the basis of a business model that both reflects an enterprise's strategic positioning in its external environment and shapes and informs its operational activities. There are potential interactions between all of the viewpoints and functional activities involved, some relating to internal enterprise activities and some to external activities and linkages. Some researchers see value-creating networks as the future of competition (Kothandaraman and Wilson [19]).

12.2 Design Thinking and Business Models

Evolving technological infrastructure is enabling business to be conceived and conducted in new ways, while broader community concerns may require the design of a new generation of business models that balance people, profit, and planet considerations. We need to think about ways to develop innovative business models to respond to these drivers. Some firms start with a business model concept, and then develop the details via an iterative learning process. Others start by clarifying the nature of their current business model, e.g., using Osterwalder and Pigneuer's [1] widely accepted business model canvas, then make incremental adjustments. It has been suggested that the application of design thinking to business models can facilitate their integrated conceptualization, definition and evolution [2], and the management literature presents a variety of ideas about how this might be achieved.

From a review of some 25 references concerned with design thinking and business models, we have made some observations:

- Concepts emerging from the design-thinking literature related to mindset (vision, a way of thinking, a focus on value innovation, balancing divergent and convergent thinking at different stages of development), process (evolutionary stages of inspiration, ideation, prototyping and implementation) and supporting tools (framing multiple viewpoints, constructing artifacts that make concepts tangible, managing configurations of artifacts).
- Requirements were represented in terms of an ontological framework associated with value, relationship, and resource architectures.
- Implementation arrangements were viewed as networks of activity systems.

In this section, we start with the application of design thinking at the concept level. We adopt a foundation analogous to that underpinning the broadly used design problem-solving process TRIZ, which evolved from the identification of common patterns in large numbers of patents. For example, some patterns related to measurement might prompt questions about how measurement could in some way help us solve our design problem.

We draw on a business typology that evolved from a review of thousands of American businesses by Malone et al [3], initially looking for patterns in relation to economic performance and business sector. They found the results inconsistent, but achieved consistent results when they considered the way business was conducted in terms of two primary components:

- The dominant type of business process (asset creation, asset trading, landlord, or deal-broker).
- The type of asset traded (financial, physical, intangible, or human). The authors had some difficulty with notions of trading human assets, and we have substituted the term intellectual assets.

Malone et al [3] gave examples of firms strategically positioned in each of the sixteen business process type/asset type domains. Drawing on practical experience, one can readily identify an example of each type, and have some broad appreciation of the nature of its value proposition and operations (e.g., a manufacturer or a real estate agent). An overview is provided in Table 12.1 with our additional business model descriptors shown in italics. This typology has a combination of transaction cost economic and resource-based views of the firm as theoretical underpinnings.

While this tabulation may seem simplistic, an enterprise may choose to move from one domain to another, e.g., move from manufacturing to retail or vice versa, combine manufacturing with retail, provide a manufacturing service only, or design and manufacture products. The Uber taxi service business blends the landlord (owner-drivers) and broker (client interface) domains. Jet engine manufacturers provide an example of multiple combinations in the so-called power-by-the-hour model which orchestrates inventor, manufacturer, physical, and financial landlord types. Popp [4] has used this typology to characterize the different IT sector business models currently in use (e.g., software vendors have online retail stores, or trade intellectual property), making the point that a business model pioneered in one industry sector can be reframed for use in another sector.

At this concept level, the questions are: what kind of business are you in; do you want to make a change; and would a particular combination introduce an element of novelty? In the next section we consider some generic design requirements for any type of business.

TABLE 12.1

Sixteen Core Business Model Types (adapted from Malone et al [3])

Trading Role	Type of Asset Involved			
	Financial	**Physical**	**Intangible**	**Intellectual**
Supplier	Entrepreneur	Manufacturer	Inventor	*Educator*
Distributor	Financial trader	Wholesaler/ retailer	IP trader	*Collaborator*
Landlord	Financial landlord	Physical landlord	Intellectual landlord	Contractor
Broker	Financial broker	Physical broker	IP broker	HR broker

12.3 Business Model Ontologies

Early management studies of business models viewed them from many different perspectives, considering particular instances or instances in a particular industry setting. In the early 2000s, Alexander Osterwalder tried to rationalize these earlier views [1], adopting a design science approach to reveal what he called a business model ontology. The outcome of this research revealed nine generic components assembled under four higher level "pillars" comprising the product offering, customer relationships, supporting infrastructure, and the value proposition. A practical spin-off from the research was the idea of a business models canvas, where the nine building blocks are presented for consideration. This has been widely used to prompt discussion within established enterprises, accumulating shared understandings of an enterprise business model and how it is implemented. Using this to represent an as-is situation, the firm may then consider how particular components could be improved.

In 2011, George and Bock [5] reported on an analysis of the business model literature viewed through an entrepreneurial lens, supplemented with a content analysis of 151 surveys of practicing managers to understand how they would conceptualize a business model. They reported the underlying requirements to be reflected in value, transitive, and resource structures. This may be aligned with the views of Osterwalder and Pigneur [1] if both the product offering and the supporting infrastructure are combined as resources, and if transitive structures are primarily associated with the market. They [1] noted that "business models are not the activities, but the structures that bound and connect a firm's core activity set to a specific set of goals." They defined "a business model as the design of organizational structures to enact a commercial opportunity." Or put another way, aligning the firm's activities with the market it is competing in. They noted that value was a common element across both practitioner and academic perspectives, that reference to resource structure associated with product, technology was most common in the practitioner dialog, and that transitive elements focused on boundary spanning and intra-organizational transactions.

Our conception of the interactions between George and Bock's three elements is shown in Table 12.2 [5]. Many encapsulations of a business model (e.g., freemium or power by the hour) are positioned in the value/value cell (A1). Gassman et al [6] outline 55 examples framed in this way. Approaches to market engagement are represented in the transaction/transaction cell (B2) and dynamic capability management (including product development) in the resource/resource cell (C3). We suggest these cells on the diagonal represent elements of strategy embedded in the firm's business model. By way of example, take the trend for banks, where the transitive orientation to

TABLE 12.2

Interactions Between Business Model Viewpoints

Viewpoint	(1) Value Structure Activity	(2) Transactional Structure Activity	(3) Resource Structure Activity
(A) Value	Understanding value structure options, e.g., how to package value propositions	Co-creating value with clients through deal-making activities	Establishing mutual understandings of value delivery expectations
(B) Transactional	Marketing the value proposition, generating revenue	Understanding transitive structures options, e.g., what to offer online or face-to-face	Activating and co-ordinating resource structures to create value
(C) Resource	Advising on value delivery capabilities, controlling costs	Providing resources supporting transaction structure activities	Understanding resource structure options, e.g., what products to offer, what to do in-house, and what to outsource

go digital means that intelligent agents interact with clients and significant supporting infrastructure has to be provided (cell B2).

From a design-thinking perspective, the ontological elements of a business model discussed may be viewed a set of generic requirements to be met:

- There must be a competitive value proposition, or no deals will be done.
- There must be targeted market actors who will benefit from accepting the value proposition, and there must be an understanding of their needs, or there may be limited transactions.
- There must be a mechanism for engaging with market actors to facilitate deal-making (e.g., via a physical or virtual front office).
- There must be a mechanism for creating and delivering value to honor deals made (e.g., via back-office functions).
- There must be a sustainable revenue and cost/benefit architecture associated with the value proposition (recognizing benefits may not be purely economic).

Figure 12.1 shows a concept diagram representing these requirements.

At the beginning of this chapter we suggested complex systems had to make operational and business sense. Our experience working with the system description framework shown in chapter 3 (Figure 3.1) and the business

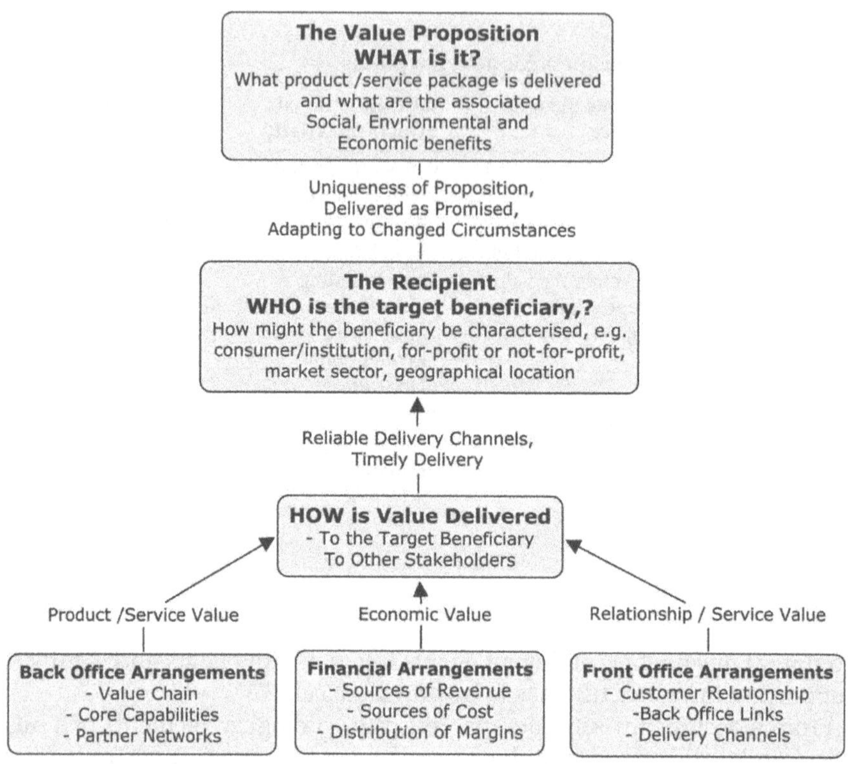

FIGURE 12.1
A business model concept diagram

model framework presented in Figure 12.1 is that they may be seen as compatible if certain conditions are met:

- Economics are expressed as a concern, and part of the system architecture rational is the establishment of a viable value proposition to address that concern.
- Customer stakeholders and/or system user stakeholders have both operational and business viewpoints to consider.
- The system architecture includes customer/user interaction functions (front office) and value creation and delivery functions (back office).

The terms ontology and business model are also seen combined at an operational level of detail in some academic literature. Gordijn et al [7] described the evolution of business model representations starting as a concept in the 1990s, then moving through phases of (a) defining and classifying business models, (b) listing business model components, (c) describing business model elements, (d) identifying reference models and ontologies, and (e) applying business model concepts using applications and conceptual tools. In pursuing an application viewpoint, they considered the nature of interactions between some of the ontology "building blocks" represented in the business model canvas. Telang and Singh [8] presented an operational-level ontology where tasks carried out to honor commitments between parties were associated with six particular interacting influence factors. These are described as agent, role, goal, task, capability, commitment (a credit/debtor transaction perspective), and business relationship. Gordijn and Akkermans [9] considered three viewpoints associated with e-commerce business models (value, business process, and system architecture) combined with the stakeholders involved, the primary focus, and methods of representation of each viewpoint. They described an "e^3-value" ontology drawing on nine particular attributes: actor, value object, value protection, value interface, value exchange, value offering, market segment, composite actor (a collaboration of sorts), and value activity. These were used to develop case maps of a particular application instance.

A common theme in these business model representation tools is the consideration of actors, activities, and their relation to particular objects.

An operating business model could be viewed as a system of internal and external activities that creates value and appropriates a share of that value. The architecture of the activity system can be described in terms of content (what activities should be undertaken), structure (how should they be linked and sequenced) and governance (who should perform them and where) in pursuit of a particular design theme (expressed as seeking novelty, lock-in, complementarities, or efficiency). It has been suggested that to explore the relationship between individual actions, organizational activities and business model performance further, scholars could also draw on activity theory, which has received scant attention in the management and organization literatures to date. This idea is pursued in the next section.

12.4 Business Models as Activity Systems

O'Leary [10] has suggested that enterprise ontologies may be examined from an activity theory perspective. He argued that activity theory, based on psychology, provides a template-based approach in capturing the context of individual activity in an organization. In particular, activity theory uses eight key class concepts: activity, outcome, subject, object, community, rules, tools, and division of labor, as a means of organizing and capturing context information. He concluded that an activity theory approach drew out

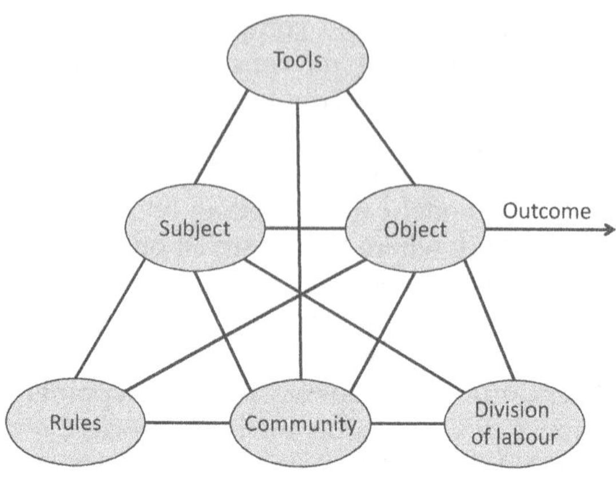

FIGURE 12.2
The activity theory framework—a model of work activity

information typically not included in enterprise ontologies, but likely to be of interest to most enterprises. The activity theoretical framework is shown in Figure 12.2.

Activity theory has been used to facilitate the study of workplace design [11], business process modeling, and business strategy deployment and development, where both the tools used and the application context were important. Jones and Holt [12] drew on an activity theory framework to study the creation and evolution of new business ventures, illustrating the contradictions and tensions that nascent entrepreneurs have to deal with. In activity theory, six generic elements interact to deliver an outcome. The core idea is that activities undertaken by a subject with an object in mind are moderated by the tools available. This was extended to include other matters of context: rules, community, and division of labor. This six-element framework has the following properties:

- Each element may be regarded as an activity system itself with its own particular object, e.g., to establish and operate a team (subject), or to develop and provide tools, drawing on the six generic element framework in a recursive way at a sublevel.
- There are two-way interactions between each element (e.g., a community may define some rules or may be influenced by some rules).
- One element may moderate actions between any other two elements (e.g., rules may moderate interactions between a subject and a community).
- There is likely to be a level of contradiction or tension between any two elements, particularly if one element changes (e.g., conflict between the desired object and the practical ability of the subject to deliver).

It has been found that activity theory may have to be interpreted in a particular way in each application of it. Blackler et al [13] described a firm's expertise as culturally situated, linguistically and technologically mediated, and socially distributed. They used an activity theory framework to map the evolution of the firm in the case study, the expertise it used, and some innovation-related activities it carried out over time, clarifying actor roles in different situations and how influence factors had changed. They noted different tensions had to be managed at different times. The aim of the firm in their case study was to enhance design as a process of problem solving, and the activity theory framework was reframed in this context.

In the context of business model design, we have suggested the following. Firstly, business models support the competitive positioning strategy of an enterprise—what kind of business are we in and how do we engage with the market? Secondly, different kinds of dynamic capabilities are needed to support this strategy. And thirdly, these considerations provide operational context for an enterprise business model where interactions between these capabilities would have to be considered in thinking about an interpretation of the activity theory framework. The need to establish a value proposition is a requirement at every level of business model analysis and will be made the object of our activity theory representation of a business model. In doing this, we also considered framing the representation in language understood by practitioners who support business model innovation.

12.4.1 Renaming and Reframing the Six Activity Theory Elements

We suggest the object element be renamed "value proposition" in a business model context. As noted earlier, a common theme linking multiple levels of business model analysis is the value proposition. Different dimensions of value suggest linkages with other activity theory elements: how to create value through products and services, how to extract value in exchange, how to deliver value through targeted use of resources, and how to facilitate value in use. We also support the idea of value co-creation with clients.

We have observed in our business model innovation casework that there needs to be a champion of the value proposition and any associated changes to it who has the kinds of competencies needed to promote it. We suggest the term "subject" be replaced with the term "service entity" as the champion/agent engaging with a client. This champion may be an individual, a group, or a real or virtual enterprise made responsible for value co-creation negotiations. The representation of a service entity is also consistent with the observation of George and Bock [5] that, from an entrepreneurial perspective, business models have a strong deal-making or opportunity-capture orientation.

In a business model innovation context, we suggest the term "tools" be replaced by "dynamic capabilities," made up of tradable assets (financial, physical, intangible, or intellectual) and infrastructure assets (financial, technological, knowledge, relational) representing the structural aspect of an enterprise. This concept of dynamic capabilities complements the resource-based view of a firm, with adaptive, absorptive, and innovation capabilities responding to market changes (or marketplace, in our model) using firm-specific processes (value networks, in our model). "The competitive advantage of firms is seen as resting on distinctive processes (ways of coordinating and combining), shaped by the firm's (specific) asset positions (such as the firm's portfolio of difficult-to-trade knowledge assets and complementary assets), and the evolution path(s) it has adopted or inherited."

We suggest the term "community" be replaced with the term "marketplace" to reflect both a business perspective and recognition that a particular enterprise may be associated with a particular professional and social community. We suggest marketplace actors include those identified by Porter [14]: customers, suppliers, competitors, potential substitutes, and new entrants as well as governments and communities of practice. They influence the nature of the enterprise value proposition and associated rules, and there are background interactions between the actors leading to change over time—another activity network.

We suggest the term "rules" be replaced with term "cost/benefit architecture." Business model studies consider both revenue and cost models and refer to financial architectures that support financial viability [15]. But there are also intangible benefits associated with a value proposition, e.g., temporal factors such as fast delivery, which can be seen as a cost or a benefit, and there may be trade-offs between the two. Over time, and in response to market factors, an enterprise will assemble a portfolio of potential benefits it can offer to clients. Costs and benefits may be associated with both the assets employed (e.g., capital utilization—a factor not generally identified in the business model literature) and marketplace expectations (e.g., regulatory compliance).

Finally, we suggest the term "distribution of labor" be replaced with the term "value network." This involves combining the enterprise value chain concept that outlines internal value-adding and value-supporting activities of an enterprise with the practice of outsourcing or strategic partnering, and recognizing the customer as a market actor who may also contribute to, and directly draw upon, the broader value network. Similarly, competitive positioning has moved beyond individual firms to value-creating networks using three core building blocks: superior customer value, core competencies, and relationships.

The resultant business model activity system is shown in Figure 12.3. The terms used in the classical activity theory framework are shown in parentheses in the figure and ensuing discussion.

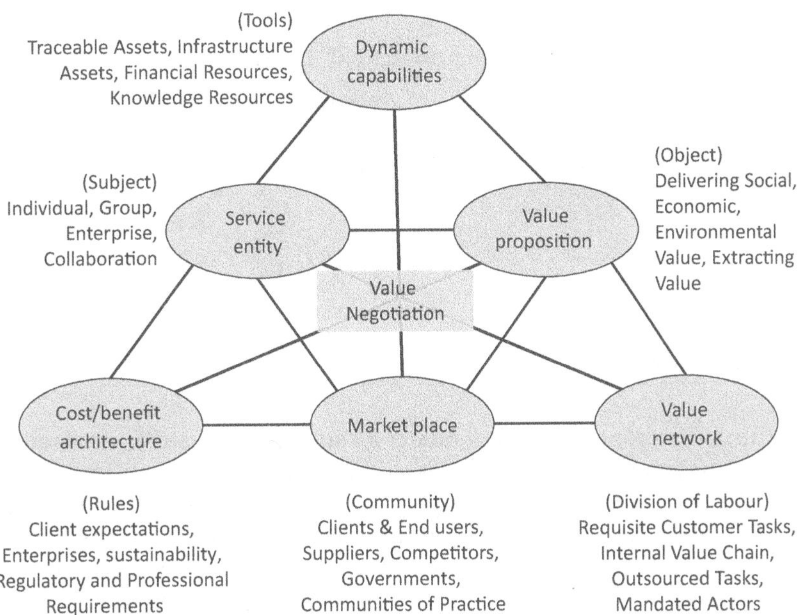

FIGURE 12.3
Activity theory framework reflecting a business model

12.4.2 Element Properties and Interactions

In the previous discussion, we observed that each element of Figure 12.3 was a complex entity itself. We noted that the activity theory framework can be used in a hierarchical way—used as a way of thinking about any activity undertaken to achieve any object at any level of analysis. For example, an activity may be undertaken with the object of enhancing enterprise dynamic capabilities (tools).

Another property of activity theory elements is that a relationship between any two individual elements is moderated by a third element. We observe that all of the elements may have a moderating effect, depending on the viewpoint. For example, in negotiating a value proposition (object), the service entity (subject) may be influenced by any combination of the dynamic capabilities (tools) available, the marketplace (community), cost/benefit architecture (rules), and the value network (division of labor).

Activity theorists also suggest the interactions between individual elements present as contradictions or tensions that may stimulate thinking about innovative solutions, e.g., tensions between the cost of a particular value proposition to a client and the cost to the service entity. This may help explain the evolutionary nature of business model development, as there are many possible interactions to consider.

TABLE 12.3

Fifteen Kinds of Interaction Within an Activity Theory System

	Interaction Element					
Reference Element	**Object**	**Subject**	**Tools**	**Rules**	**Community**	**Division of Labor**
Object		(1)	(2)	(3)	(4)	(5)
Subject			(6)	(7)	(8)	(9)
Tools				(10)	(11)	(12)
Rules					(13)	(14)
Community						(15)
Division of Labor						

In thinking about a business model as an activity system, we drew on a systems engineering tool, the interaction matrix, to map multiple interactions. The six elements can be mapped against each other, yielding 36 cells. The diagonal, e.g., marketplace-to-marketplace interaction, is normally considered null; but, as noted earlier, we interpret this in terms of interactions within that element, e.g., the interactions between marketplace actors. Interactions may be described as unidirectional or bidirectional. For example, how do dynamic capabilities influence the value proposition, and how does the value proposition influence the requisite dynamic capabilities? Alternatively, we could put both together and ask what the interactions between the value proposition and dynamic capabilities are. Taking this latter approach gives 15 combinations to consider (see Table 12.3). For example, what is the nature of interactions between the value network (division of labor) and the service entity (subject)?

Some implications of each combination are illustrated in Table 12.4 using the numerical identifiers from Table 12.3.

12.4.3 Constructing and Interpreting an Activity Theory Model

In our prior experience with reframing activity theory in a practitioner context, we have found that, once this is done, the theoretical framework makes intuitive sense to practitioners. For example, in working with small firms to establish collaborative ventures, calling the object "the deal," replacing the division of labor with "who does what," and the subject with "who will take the lead" got potential participants engaged quickly.

The terminology adopted in Figure 12.2 reflects its academic origins. In this section, we associate each of the elements in Figure 12.2 with a set of plain-language questions intended to facilitate practitioner engagement. The questions are framed in thinking about the context of each element and its linkages with other elements. We speculate that stimulating discussion with practitioners using a mix of familiar and unfamiliar terms draws them out of their traditional thinking patterns in considering innovative business models.

TABLE 12.4

Examples of Reframed Activity System Element Interactions

Interacting Activity System Elements	Example(s) of Interaction and Opportunities for Innovation
(1) Value proposition and service entity	Mechanisms for negotiating a "deal" that co-creates value with the client.
(2) Value proposition and dynamic capabilities	Providing access to tradable assets; feedback on effective resource configurations supporting value in use.
(3) Value proposition and cost/benefit architecture	Consideration of alternative cost/benefit configurations; revenue streams; value capture arrangements.
(4) Value proposition and marketplace	Communicating the offer; obtaining feedback on market valuation of the offer.
(5) Value proposition and value network	Bundling contributions from multiple sources.
(6) Service entity and dynamic capabilities	Defining requisite assets and resources; creating desirable value creation configurations.
(7) Service entity and cost/benefit architecture	"Rules for the game": profit models; desired modifications.
(8) Service entity and marketplace	Customer engagement; brand establishment; the identification of unmet needs.
(9) Service entity and value network	Market channels and alternative value creation options from combinations of internal and external functions.
(10) Dynamic capabilities and cost/benefit architecture	Identification of novel tradable asset features; common understanding of cost optimization requirements.
(11) Dynamic capabilities and marketplace	Understanding what capabilities are competitive in the marketplace and what capabilities the marketplace is looking for.
(12) Dynamic capabilities and value network	Establishing what to outsource and what to insource; obtaining leverage from collaborations.
(13) Marketplace and cost/benefit architecture	Understanding performance requirements driven by clients and competitors plus community and professional norms; developing and promoting a brand.
(14) Marketplace and value network	Nominating preferred actors by the marketplace; identifying new marketplace opportunities by the value network.
(15) Cost/benefit architecture and value network	Understanding internal and external trade-offs in cost structures; identification of value proposition benefits from partnering.

In the following, we use an original activity theory element as a key, provide commentary based on our proposed adaptation (see Figure 12.2 and Table 12.4):

- **Object**—What is the compelling deal you offer to clients, and how does it help them realize value in use? What is your value proposition? What social, environmental, and economic aspects of an offer do clients value? How is the deal finalized and monetized? What combination of tangible or intangible enterprise assets and value network connections influence deal-making?
- **Subject**—What kinds of real or virtual agent interact directly with clients (individual, group, enterprise, collaborative venture)? What value-based competencies are required by this service entity (technical skills, change disposition, conflict management, market acuity, coordinated logistics, knowledge channels, or fluid partnering)? What resources are accessed in packaging a deal, and what value network connections are mobilized to deliver value? Which kinds of interactions are face-to-face and which are online?
- **Tools**—What kind of business are you in and what kinds of dynamic capabilities have you established to support the enterprise business strategy in a changing world? What specific kinds of assets do you offer for trading (financial, physical, intangible, or intellectual) and what is your mode of trading (make for sale, retailing, landlord, or broker)? What kinds of internal and external infrastructure assets do you need to support your operations (financial, physical, knowledge, or relational)?
- **Community**—Who are your customers, suppliers and competitors, and how do they influence your value proposition? In this marketplace, do you have to consider both direct customers and end users of your product/service? Where are they located and how do local community or government actors influence market access and cultural norms? What communities of practice are you engaged with and how do they influence market access?
- **Rules**—How does your cost/benefit architecture provide a basis for sustainable trading? What do customers expect? What tangible and intangible benefits to do you offer customers and other stakeholders, and how do they support the sustainability of your enterprise? What capital assets and other resources support the delivery of these benefits? What are your sources of revenue and what are your customers prepared to pay for, and not pay for? What costs are incurred in delivering customer benefits and investing in the future? What rules have to be complied with, and at what cost?

- **Division of labor**—Describe your operational value network. What contribution to value co-creation is provided by the customer? How are internal enterprise support and value creation activities organized to deliver customer value and extract enterprise value? What tasks are outsourced to strategic suppliers/partners? Are operations based on collaborative arrangements? What tasks are mandated by contractual/regulatory requirements and do particular actors have to be engaged (e.g., auditors)?

12.5 Business Model Innovation

Question: How is that Google has become a large global player by providing a free online service and free software? (Answer: It makes a small amount of money from a large number of transactions, e.g., through advertising.) Question: Why do airline operators lease jet engines from General Electric or Rolls-Royce rather than buy them? (Answer: They make substantial savings in capital and infrastructure costs, and can focus on their primary business.) Question: How is it that you can buy a desktop printer for about the same price as a set of ink cartridges? (Answer: There is a high margin on the sale of the consumables.) These examples illustrate the enactment of business models that may not make sense at first sight, and may have their origins elsewhere. The printer example is based on a much earlier application known as the razor-blade model.

Question: What is business model innovation? The previous examples illustrate one approach—completely change the enterprise business revenue model. In a European Commission study, 60 firms around the world that claimed to have embraced business model innovation were surveyed [16]. Were they undertaking product, process, organizational, or marketing innovations—the usual categories considered in innovation surveys? The most common answer was that innovation on two or more fronts were combined. A review of the focus of innovation showed the provision of new services to be the most popular, and about 60% of the firms surveyed thought what they were doing represented a radical change. A cluster analysis indicated three kinds of innovation focus:

- Goods innovators represented the smallest group (about 20%), but showed a high level of innovation activity.
- Revenue model innovators represented about 30% of cases, and had a strong focus on service innovation.
- Small-scale innovators, the largest group, worked on fine tuning some aspect of a business model. This group rarely saw their efforts as radical.

A longitudinal study by Sonsa et al [17] followed the experience of Naturhouse, a Spanish dietary products business, which had to learn by trial and error how to best configure a product-service combination offered through service outlets, and realize growth through a franchising strategy. Two stages of exploratory activity and two stages of exploration activity were observed over the 20-year evolution of the business. The owner-manager's persistence with the business model supported continuing activity in times of hardship, and he drew on the collective knowledge of members of the organization. Over the evolutionary period, while the primary focus was on service innovation, there were also an accumulation of small-scale product and practice innovations as the business model was fine-tuned in response to both internal and external needs and opportunities. The first point to be made here is that business model innovation may be both revolutionary and evolutionary at the same time for the firm involved. The second point is that at a particular point in time, the focus of innovation may lie in any one of the nine cells shown in Table 12.2.

We now briefly discuss some of the properties of the activity theory framework identified in section 12.4 that may stimulate thinking about opportunities for innovation:

- Adopting the activity theory framework as a way of thinking, each element may be regarded as an activity system itself with a particular object. For example, if we wish to further develop our marketplace (object), who is going to do it, what resources (tools) will be needed, and what are the rules for the activity? The same questions might be asked if we want to change our cost/benefit architecture.
- Think about the two-way interactions between each element (e.g., a community may define or be influenced by some rules), and recognize there is likely to be a level of contradiction or tension between any two elements, particularly if one element changes (e.g., conflict between the desired object and the practical ability of the subject to deliver). Seek out contradictions and tensions, just as concerns were identified as drivers in chapter 3. For example, consider interaction (13) in Table 12.4, between marketplace (community) and value network. We have seen case studies of government clients (representing a market) seeking public–private partnerships (representing a value network) to develop or support complex public assets, and studies of small firms collaborating under a collective brand to establish a value network to access new markets, such as major government projects.
- Recognize that one element may moderate actions between any other two elements, creating a string of interactions to consider. For example, consideration of our cost/benefit architecture (rules) may moderate interactions between our service entity (subject) and our marketplace (community). This may emphasize an economic view.

Alternatively, If we wish to explore the interaction between our value proposition (object) and our marketplace (community), by considering non-price components of a current value proposition, the contribution of products/services (tools) is highlighted (#2 in Table 12.4), making this the moderating element. This in turn leads to exploration of how such product/services may be adapted to support unmet marketplace needs (#11 in Table 12.4).

12.6 Case Study

Our case study enterprise is Rolls-Royce. The company manufactures aircraft engines and has changed its business model over time, occasionally operating two business models in parallel. We will focus on engines for commercial airliners, although similar observations might be made in relation to some military aircraft operations. We draw on an article by Smith [18], who considered the role of technology in the evolution of the business model.

Commercial airliner manufacturers may choose to offer a choice of jet engines for their aircraft to their clients. The rationale may be that the client already has infrastructure in place for working with one brand, or small differences in engine characteristics may suit a particular mission better. Engine manufacturers compete to have their engines included in the specification, and this may put pressure on the initial cost of the engine. Engine manufactures may focus R&D on performance at the expense of higher levels of maintenance, making money on the provision of spare parts—a version of the razor-blade model mentioned earlier. They may also offer technology updates for older engines, reflecting their investments in intangible assets like patents.

Rolls-Royce first offered an alternative business proposition for a small business-jet engine in the early 1960s—the power-by-the-hour business model. A complete engine and accessory replacement service was offered on the basis of a fixed cost per flying hour. This aligned the interests of the manufacturer with that of the operator, who only paid for engines that performed well. Rolls-Royce launched a service called CorporateCare in 2002, adding a range of additional features. These included in-flight engine health monitoring; availability of leased replacement engines during off-wing maintenance, thereby minimizing downtime; and a global network of authorized maintenance centers to ensure that world-class support is readily available to customers whenever required. This service allowed operators to remove risk related to unscheduled maintenance events and make maintenance costs planned and predictable.

This change in the value proposition impacted all aspects of the operations characterized in Figure 12.3. And while the core value proposition (power by the hour) was not changed, Smith [18] points out how this business model and the evolution of technology have incrementally but dramatically changed Rolls-Royce operations over time. Whereas the earlier primary focus on performance (which is still key in military applications) and the razor-blade model did not encourage efforts to reduce the cost of maintenance and spare parts, being responsible for both the initial and operating costs of the engine stimulated research into increasing engine durability and ease of maintenance. And the business needed to operate differently. As noted in relation to Table 12.1 about generic types of businesses, the power-by-the-hour model requires orchestration of the inventor, manufacturer, physical landlord, and financial landlord types. This business model requires consideration of higher levels of capital funding by Rolls-Royce. Maintenance and support operations have to be established around the world. Systems have to be established to collect and assess massive 24/7 flows of data from engines operating around the world, and observed unsatisfactory trends or malfunctions need to stimulate appropriate actions.

Having established global operations, Rolls-Royce was also better positioned to pursue additional after-market services. Building on this experience, in 2003, the company negotiated somewhat similar performance-based contracts with the US Navy.

12.7 Reflections

The underlying mission/goal associated with the business model concept is the creation, delivery, and appropriation of something that is valued—goods, services, or a combination of the two—in a way that is economically sustainable. The environment within which these activities take place is conditioned by the capabilities of the particular enterprise and the dynamics of markets served. We have presented a three-tiered way of thinking about business models: at a conceptual level, what kind of business have we established; at a requirements level, what elements of structure have to be considered; and at an implementation level, how are these elements represented and linked? Certainly, the way business is done is impacted by the infrastructure and technology available, e.g., the internet supporting e-commerce, but our case example also illustrated how technical developments can impact the way a particular business model is implemented.

In reflecting on this chapter, we suggest you revisit section 12.2 and consider these questions:

- Can you list three or four different kinds of businesses you have observed in the context of Table 12.1?

- For each kind of business, what are the elements of structure represented in Figure 12.1?
- For each individual business, what are the specific implementation arrangements and opportunities for innovation, drawing upon ideas presented in section 12.4?

12.8 References

1. Osterwalder, A. & Pigneur, Y. *Business Model Generation*, https://strategyzer.com/books, 2009.
2. Martin, R. (2010). "Design thinking: achieving insights via the 'knowledge funnel.'" *Strategy & Leadership*, 38(2), 37–41.
3. Malone, T.W., Weill, P., Lai, R.K., D'Urso, V.T., Herman, G., Apel, T.G., & Woerner, S. (2007). "Do some business models perform better than others?" MPRA Paper No. 4752, online at http://mpra.ub.uni-muenchen.de/4752/, posted September 7, 2007.
4. Popp, K. (2011). "Software industry business models." *IEEE Software*, 28(4), 26.
5. George, G. & Bock, A.J. (2011). "The business model in practice and its implications for entrepreneurship research." *Entrepreneurship Theory and Practice*, 35(1), 83–111.
6. Gassmann, O., Frankenberger, K., & Csik, M. *The Business Model Navigator: 55 Models That Will Revolutionise Your Business*, Pearson UK, 2014.
7. Gordijn, J., Osterwalder, A., & Pigneur, Y. (2005). "Comparing two business model ontologies for designing e-business models and value constellations." *BLED 2005 Proceedings*, 15.
8. Telang, P.R. & Singh, M.P. (2012). "Specifying and verifying cross-organizational business models: An agent-oriented approach." *IEEE Transactions on Services Computing*, 5(3), 305–318.
9. Gordijn, J., Akkermans, H., & Van Vliet, H. (2000). "Business modelling is not process modelling." In *International Conference on Conceptual Modeling*, Berlin, Heidelberg: Springer, pp. 40–51.
10. O'Leary, D.E. (2010). "Enterprise ontologies: Review and an activity theory approach." *International Journal of Accounting Information Systems*, 11(4), 336–352.
11. Engeström, Y. (2000). "Activity theory as a framework for analyzing and redesigning work." *Ergonomics*, 43(7), 960–974.
12. Jones, O. & Holt, R. (2008). "The creation and evolution of new business ventures: an activity theory perspective." *Journal of Small Business and Enterprise Development*, 15(1), 51–73.
13. Blackler, F., Crump, N., & McDonald, S. (1999). "Organising processes in complex activity networks." *Organisation*, 7(2), 277–300.
14. Porter, M.E. (2008). "How competitive forces shape strategy." *Harvard Business Review*, January, pp. 25–41.
15. Teece, D.J. (2010). "Business models, business strategy and innovation." *Long Range Planning*, 43(2), 172–194

16. Chesbrough, H. (2010). "Business model innovation: opportunities and barriers. *Long Range Planning*, 43(2), 354–363.
17. Sonsa, M., Trevinyo-Rodríguez, R.N., & Velamuri, S.R. (2010). "Business model innovation through trial-and-error learning: The Naturhouse case." *Long Range Planning*, 43(2), 383–407.
18. Smith, D.J. (2013). "Power-by-the-hour: the role of technology in reshaping business strategy at Rolls-Royce." *Technology analysis & strategic management*, 25(8), 987–1007.
19. Kothandaraman, P. & Wilson, D.T. (2001). The future of competition: value-creating networks. *Industrial Marketing Management, 30*(4), 379–389.

12.9 Additional Reading

Amit, R. & Zott, C. (2015). "Crafting business architecture: The antecedents of business model design." *Strategic Entrepreneurship Journal*, 9(4), 331–350.

13

Innovation in Ecosystems

> Innovation is always a surprise. By definition, it is something no-one has thought of before. Its very existence shows that reality is not fixed in predictable patterns. Instead creative new possibilities can emerge in any field, in any industry. Innovators see new patterns in the familiar, apparently immutable, situations. It is though they see the world through a kaleidoscope, which creates endless variation from the same set of fragments.
>
> **Rosabeth Moss Kanter, 2002**

13.1 The Centrality of Ideas and Champions

As noted at the start of this chapter, ideas may emerge from the perception of new patterns of existing things, from a need for change, or from structured processes of experimentation. Still, as 1937 Nobel Prize winner Albert Szent-Györgi is quoted as saying, "Discovery consists in seeing what everyone else has seen and thinking what no one else has thought." In addition, sensing the right time and place to promote an idea has been associated with intuition and entrepreneurship [1]. Sarason et al [2] see that "entrepreneurship is presented as the nexus of opportunity and agency, whereby opportunities are not single phenomena, but are idiosyncratic to the individual." In other words, entrepreneurs see opportunities others may not see.

It is recognized that without a champion who shows persistence, belief, and commitment to an idea, an innovation may not realize its full potential (Schon [3]). Coakes and Smith [4] suggest that "championing innovation must become a norm in organizations and not an episodic event that relies on happenstance and a strong-minded individual expending large amounts of effort." Support for both an individual idea and the process of innovating is required.

13.1.1 Innovation Champion Roles

Some researchers observed that different kinds of champions helped overcome implementation problems in different stages of the evolution of an

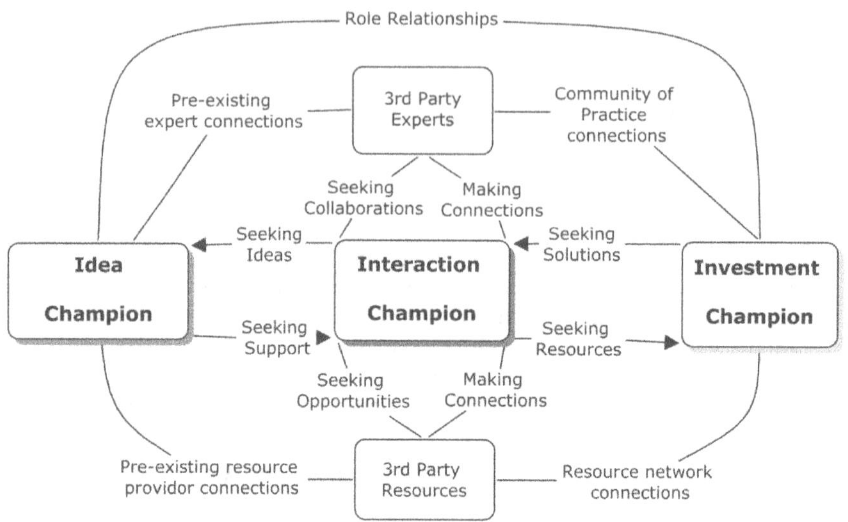

FIGURE 13.1
A view of innovation champion interactions

innovative product or process [3]. A multiple champion perspective can identify the types of actors need to be present on the innovation platform. During the initial stages of the evolution of an innovation, it is normal that individual champions should fill multiple roles.

In our research studies, we have observed more than 25 role descriptors that could be associated with the notion of an innovation champion. Following the cue from other authors (e.g., Orlikowski et al [5]) to cluster roles under a set of meta-level functions, we suggest there are three innovation champion meta-roles to be enacted, each supporting a different aspect of innovation. Those with innovative ideas need investment to develop and deploy them. Those making strategic investments in innovation need a flow of ideas to realize benefits from the resources they have allocated. In between there are those that facilitate the process of innovating by making connections with internal and external political supporters and knowledge or infrastructure resource providers. A simplified view of the interactions between these roles is shown in Figure 13.1. We subsequently explore the question: How do champion roles that stimulate and support innovation interact?

13.1.2 Innovation Champion Activities

Structuration theory (Giddens [6]) has been used to help understand the interaction between actors and structure in an innovation context. In this theory, rules (methodologies) may appear to exist independently, but they

TABLE 13.1

Practices Adopted by Innovation Champions

	Innovation Champion: Elements of Agency		
Elements of Social Structure & Modality	**Investment Champion (promoting innovation as a strategy)**	**Interaction Champion (facilitating the practice of innovating)**	**Idea Champion (promoting an individual innovation)**
Signification: structures that help produce meaning through interpretive schemas and communication. Innovation must make sense.	Pronounces the adoption of innovation as a strategic management tool, launches programs to stimulate and facilitate innovation.	Embraces an idea and communicates widely within a work unit, across work units, and outside the host organization to facilitate its evolution.	Enunciates an idea, refines a concept, imagines the potential future impact of an innovation.
Legitimation: structures drawing on social norms, values, and standards that are sanctioned. Innovation must be embraced/endorsed.	Establishes an innovation culture with supporting permissions and procedures, e.g., supporting open or collaborative innovation practices.	An individual or organization gaining respect/trust through their actions and/or via their appointment by an investment champion.	Demonstrates the utility of an idea, demonstrating organizational fit.
Domination: structures that produce and exercise power through the allocation of physical resources (allocative) and human resources (authoritative). Innovation requires access to diverse resources.	Establishes infrastructure to facilitate idea discovery, development, and deployment. Assembles and authorizes the use of resources.	Persists under adversity. Uses "know-who" knowledge as power and facilitates access to third-party knowledge and resources.	Uses technological and/or application knowledge as power and stimulates multi-disciplinary teams to enact an idea.

are only applied through use and reproduction in practice, where their persistence may be framed as "culture." Giddens identified three types of structure as important, each associated with a particular interaction modality. In Table 13.1 we have mapped our three innovation champion meta-roles against these three types of structure, seeking to situate the multiplicity of functions described in the literature.

13.1.3 The Adaptable Innovation Champion

As an idea evolves, different professional communities contribute to its development and application. For example, at the early stages, technical functions are responsible for scanning the technology environment, imagining possibilities for emergent technologies, while managerial functions are responsible of scanning the market environment, imagining possibilities in emergent client needs and markets, or "picking winners." Handovers between may be necessary between such organizational groups, e.g., from research to engineering development. If a person or group does not effectively pick up the each of the three champion meta-roles, outcomes may be suboptimal. In smaller enterprises, we have observed that a particular individual may take on different roles at different times. For example, a manager who is the investment champion in the early stages may become the idea champion at a later stage to promote the innovation outside of the enterprise.

13.1.4 Mapping Actor Connections Using Social Network Analysis

Coulon [7] undertook a literature review of the use of Social Network Analysis (SNA) in innovation research, observing an increasing interest in the topic since 2000. In undertaking SNA, researchers need to decide on the unit of study (e.g., a family or a region), on the relational form and on the content to be studied. We have studied individual innovation projects concerned with the definition and development of novel ideas to at least the proof-of-concept stage. The relational form explored was structural with content ties being classified according to Knoke and Yang [8]:

- Communication relations—Linkages between actors as channels through which messages may be transmitted.
- Transaction relations—Actors exchange control over physical or symbolic media, for example, in gift-giving or economic sales and purchases.
- Authority/power relations—These types of ties, usually occurring in formal hierarchical organizations, indicate the rights and obligations of actors to issue and obey commands.
- Instrumental relations—Actors contact one another in efforts to secure valuable goods, services, or information, such as a job, political advice, or recruitment to a common cause.

Three analyses were typically undertaken for each project—one around project formation, one characterizing the development stage, and the third at the deployment stage. We had previously observed that if all phases were pooled together, the nature of handovers and role changes was masked. As we assembled lists of contributing project actors, influential actors outside the project team were identified. An individual project's network of accessible contributors grew through contacts that could be made via external sponsors, lead users, and cross-sector industry linkages. This highlighted the fact that the project was not isolated from a broader innovation ecosystem.

13.2 Regional Innovation Systems

Adner and Kapoor [9] argue that the success of an innovating firm often depends on the efforts of other innovators in its environment. They present a model suggesting that a particular firm may rely on different kinds of components from other firms to develop and deploy their product, while customers may need to bundle this product with other complements to obtain value from use of the product. Components and complements may themselves emerge from the innovative efforts of others. By way of example, the uptake of electric cars may be inhibited by battery capacity limitations and potential issues with the time and place of recharging. Autio and Thomas [10] suggest that a coherent understanding of ecosystem creation requires a multi-theoretic approach, combined with an appreciation of three related architectures—the technological architecture, the activity architecture, and the value architecture arising from the interaction between technological and activity architectures. We observe that these architectures differ from place to place.

13.2.1 Innovation Hot Spots

The World Intellectual Property Organization analyzed patent filings at different location around the world covering the period 2011 to 2015. Ten innovation hot spots were identified and ranked:

1. Tokyo-Yokohama, Japan, with an orientation toward electrical machinery, apparatus, and energy
2. Shenzhen-Hong Kong, China, with an orientation toward digital communication
3. San Jose–San Francisco, California, USA with an orientation toward computer technology
4. Seoul, Republic of Korea, with an orientation toward digital communication
5. Osaka-Kobe-Kyoto, Japan, with an orientation toward electrical machinery, apparatus, and energy
6. San Diego, California, USA with an orientation toward digital communication
7. Beijing, China, with an orientation toward digital communication

8. Boston-Cambridge, Massachusetts, USA, with an orientation toward pharmaceuticals
9. Nagoya, Japan, with an orientation toward transport
10. Paris, France, with an orientation toward transport

Clarysse et al [11] have observed that there may be a gap between the knowledge available in a region and the capability of regional businesses to extract value from it, which may be interpreted as a flaw in the activity architecture referred to earlier. Therefore, to transform a region into an innovation ecosystem, it is necessary that different actors (e.g., government, industry, and academia) should collaborate and co-create value.

13.2.2 Characterizing National Innovation Systems

Beckett and Zhang [12] have suggested that a nation's innovation system influences and is influenced by a number of factors:

- The nature of enterprises within it: for-profit or not-for-profit, commercial or government, large or small, and regional, national or global scale.
- The nature of dominant markets: market sectors (e.g. agriculture or manufacturing), market lifecycle stage, and macroeconomic conditions.
- The nature of government policies and programs relating to innovation, national research priorities, trade-related policies, and market failure interventions.
- The scale and scope of national infrastructure related to workforce competencies, finance, information and communications technologies, transport, and R&D capacity.
- Intermediary organizations that facilitate interaction between government, infrastructure, industry, and markets. Examples include industry associations, standards authorities, and cooperative research centers.

There are complex linkages between such elements of an innovation system, as illustrated in Figure 13.2. Interactions between the industry element of a national innovation system and the infrastructure element is the most common viewpoint taken by innovation management researchers. In this model, industry structure is directly influenced by access to educational and R&D infrastructure, or indirectly via intermediary organizations. These organizations may be established by industry, government, or researcher initiatives to either facilitate industry capability development or access to markets or both. There are also indirect influences from ICT and logistics infrastructure via the market element.

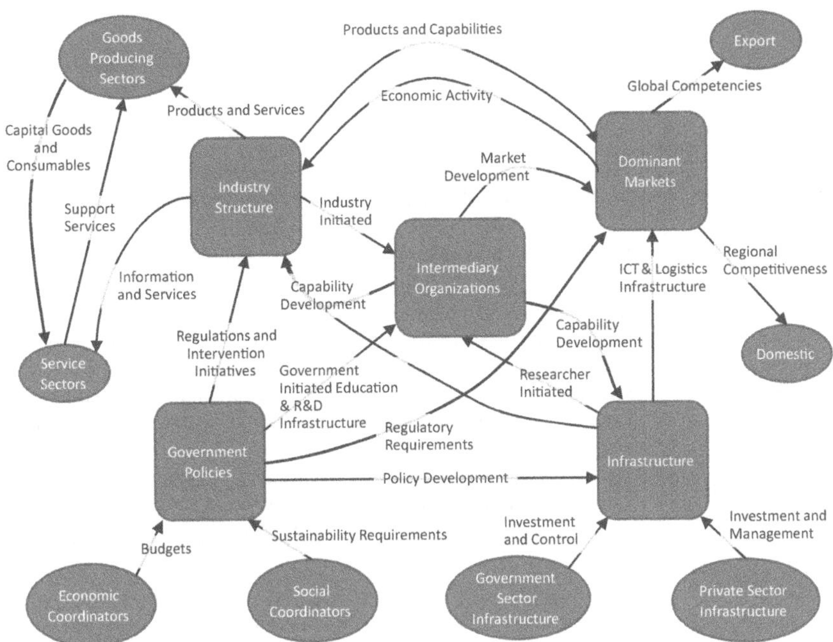

FIGURE 13.2
A representation of a regional innovation ecosystem—elements of a national innovation system

13.3 Innovation as an Evolutionary Process

As observed by March [13], delivering an innovation requires a combination of exploring possibilities where a divergent thinking mindset is appropriate and exploiting opportunities where a convergent thinking mindset is most appropriate. Building on the work of Usher [14], Garud et al [15] have noted that most innovation management researchers investigate the process of innovating as a structured series of stages. But they suggest a different view: that innovation is a complex adaptive learning process stimulated by pivotal moments—acts of insight followed by the synthesis of accumulated knowledge. They also discuss some temporal considerations, suggesting that "multiple time scales generate asynchronies across the different evolutionary stages," citing mismatches between the emergence of different components of an innovation and the infrastructure required for its development and implementation.

We suggest this is reflected in practice in various ways—in the development and testing of prototypes, and in the adoption of agile project

TABLE 13.2

Generic Stages in the Lifecycle of Innovation and Generic Transitional Events

Transitional or Lifecycle Stage	Activities and Focus
Motivation (t_0)	Starting the innovation process—"selling" innovation as a developmental strategy within or outside the innovator's world. This is the entry point.
Discovery	Searching for novel ideas or searching for solutions to problems, creatively working in the imagination. Here the *idea* is dominant.
Promotion (t_1)	Moving from searching to selecting—selling the idea within or outside the innovator's world. This is an exit point for innovators who licence or sell their idea.
Development	Selecting options for evaluation and experimentation, or refining an idea. Here creatively working the accessible *resources* is a dominant theme.
Engagement (t_2)	Moving from selecting to implementing—selling the product within or outside the innovation creator's world. This is an exit point for innovators and entrepreneurs who licence a product or sell a start-up business.
Deployment	Implementing an idea—launching it into its application domain. Here the market/application *place* is a dominant theme, creatively considering virtual, physical, and geographical possibilities.
Expansion (t_3)	Moving from implementation to maximizing value—selling the value proposition. This is an exit point for entrepreneurs who sell an established business.
Domination, or upscaling	Capturing benefits from the impact of an innovation in a competitive environment requires ongoing innovation. This is where *timing* is a dominant theme, creatively blending matters of infrastructure maturity, market readiness to support a clear value proposition, active waiting, and fast deployment.
Moving on (t_4)	This is the norm for serial entrepreneurs or project-based enterprises, which focus on ideas for identifying substitutes or making the original idea and associated capabilities a foundational component of a new initiative.
Displacement	Capitalizing on established assets and redeploying resources. Examples are the reuse of existing components in a new car model, and the successive displacement of telecommunications technologies by others that build on the established infrastructure.

management methodologies where what is learned at one stage may lead to a return to an earlier stage. Transitional events—deciding what to proceed with and how—are an essential feature of the evolutionary process. Table 13.2 presents an amalgamation of the generic stages in the lifecycle of an innovation and generic transitional events. This representation supports the potential use of Petri nets (see chapter 15) in representing innovation as a process of progressive but nested activities within discrete event-processing environments. Petri net transitions are shown in parentheses.

13.4 Matters of Time and Place

Bowen et al [16] suggest that in managing innovation, "timing is everything." The success or failure of start-up companies depends on circumstances, i.e., being in the right place at the right time has a greater influence on success than having a great idea/great team combination.

The Ancient Greeks described two notions of time. The first is chronos—linear, divisible time that we use as a management tool for coordinating activities. We focus on precise time intervals, but the concept has its foundation in cyclical astronomical events. Days are associated with the spinning of the earth and years with the journey of the earth around the sun. The other Ancient Greek notion of time—kairos—is about events in time stimulating moments of enlightenment and timely action. It is less frequently used as a management tool, but we explore its utility here. One example of a kairotic moment given in discussing the concept is injecting a game-changing thought into a debate at just the right time. Another is the moment a hunter releases an arrow, having positioned himself to access his prey and having the right tools and the skills to use them. Coessens [17] refers to kairos as framing matters of timing in an artistic performance linked in response to the background dynamic environment. But if the orator has not framed the idea, or the hunter does not have the appropriate equipment, or the performer the does not have the requisite skill, there is no impact from simply being in the right place at the right time.

When considering what contextual factors condition the ability to explore and exploit windows of opportunity for innovation, we suggest the following:

- Viewing influential events in time (in the past or predicted in the future) as windows of opportunity can provide insights into innovation evolutionary pathways.
- Four primary factors frame transitional event context that in turn shapes the most appropriate course of action. These are: place, idea, resources and time. There are multiple interactions between these elements, as illustrated in Figure 13.3.

It may be noted in Figure 13.3 that each of the four contextual factors has three sub-factors. In our experience, reflecting on their interaction (potentially as a 12 × 12 matrix) helps to consider the complexities associated with innovation, e.g., the current lifecycle stage of a particular aspect of technological infrastructure or knowledge base needed to progress a particular idea.

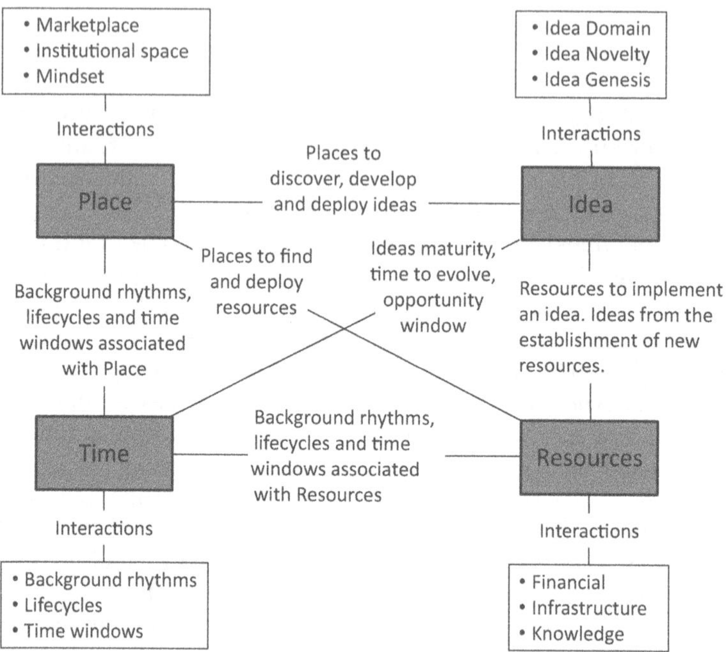

FIGURE 13.3
Contextual factors influencing timely decision-making, drawing on the Ancient Greek notion of kairos

13.5 Case Study

This innovation ecosystem case study considers the US Small Business Innovation Research (SBIR) program, which was initiated by Congress in the early 1980s. We draw on a 2003 review of the program by Audretsch [18] and our own experience with a regional government program seeking to emulate it.

In this case study, the program is discussed with reference to the system architecture framework presented in chapter 3. The mission was to respond to a perceived loss of American competitiveness in global markets. The stakeholders were the broader community, small businesses, and the government (representing the dominant markets, industry structure, and government policy elements, respectively, in Figure 13.2). The rationale here was that the majority of new jobs in the United States are created by small businesses. At the requirements level, the rationale was that particular innovations support the mission of individual agencies in serving their community stakeholders.

Consistent with some of the enterprise models presented in chapter 2, architecture descriptions are provided at concept, requirements, and implementation levels.

At the concept level, Congress mandated that each participating federal agency allocate around 4% of its annual research budget to fund innovative small firms. The program built on prior interventions by different government departments.

At the requirements level, each agency ran its own implementation of the program, generally reflecting on currently unmet needs in the community it served and future trends to inform calls for proposals from industry. At this stage, the idea champions were those government employees who had identified such needs. At this level, project proposals were reviewed for technical merit and feasibility, with selected ideas receiving phase I funding for a more comprehensive development of the idea and its potential utility. At this early stage (within the first six months) the idea champions were those putting proposals forward and those selecting them for further progression.

Phase II supported idea development, and transition to this 24-month phase took scientific/technical merit, the expected value to the funding agency, company capability, and commercial potential into account. About 40% of the initial awards progressed to phase II. Phase III funding was generally subject to the infusion of private sector investment and the uptake of the product or system in the commercial market. The original government agency idea champion sometimes became a lead user, and the investment champion role moved to the private sector. Investment decisions were informed by considering the factors shown in Figure 13.3, particularly at the transition points between stages.

At the project level, the mission became the implementation of an idea to meet a current or anticipated community need. The stakeholders were a particular agency, community, business, and investor(s). The rationale was based on delivering and extracting value from the particular idea, which had its own architectural description. An example of an individual project launched under the umbrella of a copy-cat program in Australia follows.

The project addressed the concern of a government employee in a children's hospital that conventional nipples were not effective for bottle-feeding premature infants. Following a call for proposals by a governmental industry-development agency emulating the SBIR concept, a local design-and-development firm was contracted to develop a suitable product. It was initially thought that conventional nipples were too large, and too stiff. After some trials, it was observed that breathing synchronization of premature infants during bottle-feeding was different from that of more mature infants—so the experts had learned something new, and went back to the drawing board. After some iterations, a suitable design (or architecture) emerged and was patented. The design-and-development company established a spin-off firm to manufacture the nipples in accordance with regulatory requirements (a change of place and resources in the context of Figure 13.3). During development, the firm drew on inputs (or viewpoints) from materials researchers and a polymer product manufacturer to fine-tune product attributes and economic methods of manufacture. The original idea champion at the children's hospital became the product lead user, and shared the experience gained with other hospitals, helping to speed up market development. This situation encouraged investors to support the growth of the firm in pursuing global markets (resource and place considerations, in the context of Figure 13.3).

13.6 Reflections

The word "innovation" is used in two contexts—one in relation to a novel application developed from an idea, and the other in relation to the process of developing and deploying an idea. Whilst innovation starts with an idea which may or may not be related to a new technology, there must also be some enthusiasm for the idea from an innovation champion who will promote its implementation. But matters of context are also important. "Right time, right place, right people" equals success. "Wrong time, wrong place, wrong people" equals most of actual human history.

Innovation is associated with an element of uncertainty—can we economically develop our idea and will the intended users adopt it? For this reason, some researchers [19] view the process of innovation like a complex adaptive system that is supported (or otherwise) by a broader innovation ecosystem. Our case study considers an example of support arrangements stimulating the process of small-business innovation and a particular example of an innovation developed in this setting.

Can you think of an innovation you have found particularly interesting and another that failed? Consider these questions:

- Who were the innovation champions that facilitated its evolution?
- Can you associate timelines with the lifecycles of those particular innovations (with reference to Table 13.2)?
- Were some particular combinations of time and place important in its evolution?
- What kinds of infrastructure were needed to support its evolution, and were there difficulties due to the lack of suitable infrastructure at some point?

13.7 References

1. Bhide, A. (1994). "How entrepreneurs craft strategies that work." *Harvard Business Review,* March-April issue, pp 150–161.
2. Saranson, Y., Dean, T., & Dillard, J.F. (2006). "Entrepreneurship as the nexus of individual and opportunity: A structuration view." *Journal of Business Venturing,* 21, 286–305.
3. Schon, D.A. (1963). "Champions for radical new inventions." *Harvard Business Review,* March-April, pp. 77–86.
4. Coakes, E. & Smith, P. (2007). "Developing communities of innovation by identifying innovation champions." *The Learning Organization,* 14(1), 74–85.
5. Orlikowski, W.J., Yates, J., Okamura, K., & Fujimoto, M. (1995). "Shaping electronic communication: the metastructuring of technology in the context of use." *Organization Science,* 6(4), 423–444.
6. Giddens, A. (1984). *The Constitution of Society: Outline of the Theory of Structuration.* Berkeley, CA: University of California Press.
7. Coulon, F. (2005). "The use of social network analysis in innovation research: A literature review." Lund University. http://gent.uab.cat/diego_prior/sites/gent.uab.cat.diego_prior/files/Article_to_comment_10_bis.pdf
8. Knoke, D. & Yang, S. *Social Network Analysis.* Thousand Oaks, CA: Sage Publications, 2008. ISBN: 9781412985864
9. Adner, R. & Kapoor, R. (2010). "Value creation in innovation ecosystems: How the structure of technological interdependence affects firm performance in new technology generations." *Strategic Management Journal,* 31(3), 306–333.
10. Autio, E. & Thomas, L. (2014). "Innovation ecosystems." In *The Oxford Handbook of Innovation Management,* Oxford University Press, pp. 204–288.
11. Clarysse, B., Wright, M., Bruneel, J., & Mahajan, A. (2014). "Creating value in ecosystems: Crossing the chasm between knowledge and business ecosystems." *Research Policy,* 43(7), 1164–1176.
12. Beckett, R.C. & Zhang, G. (2010). "Practicing Innovation within National Innovation Systems: An Australian Perspective." In *Proceedings of the 11th International CINet Conference, "Practicing Innovation in Times of Discontinuity",* Zurich, Switzerland, Sept 5–7. ISBN: 978-90-77360-13-2.
13. March, J.G. (1991). "Exploration and exploitation in organizational learning." *Organization Science,* 2(1), 71–87.
14. Usher, A.P. *A History of Mechanical Inventions, Revised ed.* Cambridge, MA: Harvard University Press, 1954.
15. Garud, R., Gehman, J., Kumaraswamy, A., & Tuertscher, P. (2016). "From the process of innovation to innovation as process." In *Sage Handbook of Process Organisation Studies.* Sage Publications, London, Chapter 28, 451–465.
16. Bowen, F.E., Rostami, M., & Steel, P. (2010). "Timing is everything: A meta-analysis of the relationships between organizational performance and innovation." *Journal of Business Research,* 63(11), 1179–1185.
17. Coessens, K. (2009). "Musical Performance and 'Kairos': Exploring the Time and Space of Artistic Resonance." *International Review of the Aesthetics and Sociology of Music,* 40(2), 269–281.

18. Audretsch, D.B. (2003). "Standing on the Shoulders of Midgets: The U.S. Small Business Innovation Research Program (SBIR)". *Small Business Economics* **20**: 129–135.
19. McCarthy, I.P., Tsinopoulos, C., Allen, P., & Rose-Anderssen, C. (2006). New product development as a complex adaptive system of decisions. *Journal of Product Innovation Management, 23*(5), 437–456.

13.8 Additional Reading

Beckett, R.C. & O'Loughlin, A. (2016). "The impact of timing in innovation management." *Journal of Innovation Management*, 4(3), 32–64.

Ferreira, J.J., Dana, L.-P., & Ratten, V. *Knowledge Spillover-based Strategic Entrepreneurship*, Routledge, 2017. ISBN: 978-1-138-95074-0.

Part 4

System Operation and Configuration Management

14

Capability and Performance Assessment

> I often say that when you can measure what you are speaking about, and express it in numbers, you know something about it; but when you cannot measure it, when you cannot express it in numbers, your knowledge is of a meagre and unsatisfactory kind; it may be the beginning of knowledge, but you have scarcely, in your thoughts, advanced to the stage of science, whatever the matter may be.
>
> **Lord Kelvin**
> *Lecture on Electrical Units of Measurement, 1883*

14.1 Why Assess Capability and Performance?

A recent trend around the world among the owners of complex engineering systems such as aircraft or telecommunication systems is to consider the whole-of-life ownership requirements. This has two implications. Firstly, the value of the asset is not only reflected by the purchased price, but also from the benefits of using the system throughout its service life. The value of benefits is not easily measurable. Secondly, the difficulty of measuring the value of benefits in use leads to the concept of measuring the performance of the system. If the system performs well, it is delivering close to its value of use. If not, its perceived value may be high but the actual value is disappointing. Hence, the ability of an enterprise to extract as much value of use as possible from an asset does not simply depend on the functions of the asset. It depends also on the capability of the relevant enterprise, irrespective of whether it is a company, a consortium, a virtual enterprise, or a community. It should be noted that the asset is a system as normally understood in engineering sense, while the enterprise together with other assets coordinating with the current asset will form a system of systems for a coordinated goal.

The goal of assessing the capability of a system of systems is to determine capability levels that can be measured as a performance outcome of systems in service. The term "capability" is commonly used in the defense sector to describe the capacity or ability of defense to achieve a particular operational effect. This capacity or ability can be realized in many forms. The most common form is the possession of suitable equipment operated by a team of trained operators with the ability to perform a defined set of task under specified operational constraints. This meaning can also be applied in a non-defense sector. As the "capability" concept includes operational values, PBC style of contractual framework is applicable. There are two parts to this type of contractual framework. The first is the agreement to acquire the system. This is the traditional procurement contract that specifies functional and performance requirements of the final system. Added on to this is the sustainment agreement, a second contract that specifies the outcomes and performance requirements for in-service support. Studies show that quantifying the effect of factors such as environmental issues, safety, reliability, availability, and logistics into performance indicators is necessary for successful acquisition projects [1].

From the supplier's point of view, this shift in the system acquisition process means longer, more assured revenue streams based on long-term support and ongoing development instead of a series of big "must-win" procurements. It also opens up opportunities for service providers who are not original equipment manufacturers, to gain business by servicing existing equipment. In other words, the PBC paradigm shifts the business model toward systems of systems.

Complex engineering solutions like systems of systems are capability developments that require a long-term sustainment strategy to be considered as early as possible during system development. The increasing use of PBC and KPIs for this type of procurement demands a better understanding of this new business model.

Performance and productivity-based contract (PPBC) is a variation of PBC, in which productivity gain is an additional deciding factor. In this chapter, the terms PBC and PPBC are used interchangeably for the same type of service contract.

For a PBC contract to be effective, both parties to the contract must have a clear understanding of the expected outcomes and how they should be measured. If they are determined and implemented in a collaborative way, KPIs can play a major role in the successful outcome of a PBC project. This is as true for commercial projects as it is for defense contracts.

14.2 The Nature of Performance-Based Contracting

There are a lot of uncertainties in the PBC environment and questions have to be asked by both sides of the contract. On the one side, the contractor (the manufacturer of the system of systems or the service provider, depending on the type of contract) should ask whether it is beneficial (for the contractor) to take on the risks of the agreed commitment. The asset owner should ask whether the contractor has given sufficient evidence that they can meet the stringent conditions of the PBC environment. In all instances, the contractor is required to demonstrate to the asset owner that the product or service provided offers value for money. Of course, the lowest price is not necessarily the best value for money. If the product or service is deemed fit for purpose, but it is not performing as it should be, it is still not delivering the value for the money invested in it. All this points to one fundamental question: how will performance be assessed?

There is no one-size-fits-all approach for performance-based contracting. Each PBC contract is unique, tailored to provide each customer with a specific set of outcomes. For complex projects, a great deal of effort is required to ensure that outcomes are measurable and achievable. The hand-crafting of the PBC should be part of the design process, required to meet performance measures through clearer specification, structured maintenance, and appropriate support infrastructure.

14.3 Performance-Based Contracting Case Studies

The next question is whether the measurements that are developed by the customer and agreed to by the contractor will actually achieve the desired outcomes. To better understand the issues, we will present some case studies to explore the critical factors influencing the PBC business environment. The case studies highlight critical factors in PBC and investigate how industry adapts to the new business environment. It is clear from these case studies that the support service system will need to handle more complex under PBC.

14.3.1 JP 2008 Phase 3F

The JP 2008 3F case study concerns an Australian defense contract for the country's C4ISR capability (see chapter 2). The prime contractor is BAE Systems Australia (BAE). The contract is to provide and support a system to integrate an international satellite communications system consisting of space vehicles of multiple types, control terminals, facilities, and user terminals to provide the communication spectrum for defense purposes.

The WGS (Wideband Global SATCOM) system consists of up to seven commercially launched and strategically placed satellites. The Australian Defence Force (ADF), through a partnership with the US government, funded the sixth satellite in the WGS constellation. The WGS system provides global SATCOM coverage to the ADF with a communications capacity orders of magnitude greater than that provided by the original SATCOM system. The upgraded Australian Defence satellite communications capability (ADSCC) ground segment will be operated and maintained through a performance-based support contract for the system Life of Type (LOT) of 15 years. BAE will provide operating, engineering, maintenance, supply, training, and facilities support for the ADSCC ground segment.

Key performance requirements based on aggregate link availability and reliability (link performance) concepts will be monitored, and this will affect contract payments during in-service support. A major factor during this phase of the contract will be BAE ownership of spares for the ADSCC ground segment. As the support contractor, BAE is responsible for designing and maintaining the ADSCC ground Segment to meet and maintain all system performance requirements while establishing a cost-effective and responsive support infrastructure through the LOT of the system.

14.3.2 Over-the-Horizon Radar

Australia's over-the-horizon radar (OTHR) capability consists of three radar systems [2]. Development of the OTHR system began in the 1970s, and the original developmental radar (R3) in Jindalee became operational in 1993. Two further radar systems (R1 and R2) were added to create the Jindalee Operational Radar Network (JORN), which was delivered in 2003.

The 2010 launch of Defense's strategic reform program (SRP) was a key milestone in the history of the contract. The SRP initiative aimed to realize savings across Defense's operational and support activities in order to help fund the acquisition of future air (JSF) and naval (new submarines) platforms. OTHR was one of the first capabilities subject to a PBC with a savings target of around 20% to be achieved within a 5-year period.

Under the PBC, BAE had responsibilities to cover maintenance, logistics, engineering development, integration, installation, verification and validation, operator training, and support to the ongoing operation and enhancement of R3 and the greater JORN.

The constraint of the OTHR PBC was that engineering and maintenance systems are governed by the Director General Technical Airworthiness who specifies that all requirements be in accordance with the Technical Airworthiness Maintenance Manual.

14.3.3 Lead-In Fighter

The Australian government acquired the Hawk Lead-In Fighter for Royal Australian Air Force (RAAF) introductory fighter pilot training and other ADF air support tasks that were previously provided by other aircraft. Introductory fighter pilot training prepares pilots to confidently manage high-performance aircraft and advanced weapons technology in a lead-in fighter (LIF) type that can be employed more economically than a frontline combat aircraft.

BAE was awarded a PBC by the government to support the RAAF's fleet of Hawk LIF fast jet training aircraft. The contractor, i.e. BAE, was tasked to provide these deliverables:

- Maintenance at two operating bases
- Engineering services for management of the aircraft to military airworthiness standards
- Logistics and supply support

Contract performance is measured by the number of mission-capable aircraft available to meet the RAAF's daily flying requirements with liquidated damages paid per aircraft per day if the requirement is not met.

This contract is of interest because of the extent and skill of the engineering and maintenance operations, the nature of the availability contract, and the nature of the relationships between contractor and customer personnel.

The Hawk LIF aircraft are maintained by a two-tier maintenance organization, with operational maintenance being performed by RAAF personnel and deeper maintenance, e.g., engine overhaul, performed by the contractor. The contractor is also responsible for providing the supply support, weapon system logistics management, mission planning system computer support, software support, and training.

The objectives of the contract are to achieve the following outcomes:

- To meet the requirements for Hawk LIF weapon system preparedness through specified levels of support.
- To set up, operate and maintain the contractor's elements of the Hawk LIF support system while minimizing cost of ownership.

This case study shows there can be several output measures of contract performance. The principal measure is the availability of a contracted number of mission-capable aircraft to meet the RAAF's daily flying program requirements. The contract is relevant here because of the extent of the skilled engineering and maintenance operations, the nature of the availability measures, and the nature of the relationships between the stakeholders—contractor, customer, and the independent military airworthiness regulator. This was an early PBC project established in a "product-thinking" environment within both the customer and contractor communities, and has had to consciously incorporate customer-oriented thinking.

14.3.4 Non-Defense Performance-Based Contracting

PBC is also used for highway maintenance in Canada, the UK, Australia, New Zealand, and Finland, and to an increasing degree in other countries, including the United States. The states of Virginia, Texas, and Florida lead the way in this area, with transportation agencies facing growing needs and limited resources to maintain the highway network.

The resulting challenges have motivated transportation agencies to expand the use of contracting. This is achieved by outsourcing capabilities such as pavement marking and roadside vegetation control to contractors and measuring their performance. Moreover, transportation agencies in North America and around the world have developed a variety of methods for undertaking performance-based maintenance contracts.

In Australia, the New South Wales (NSW) government conducted a comparative maintenance study of two pilot projects of a 100-kilometer stretch of road in 1990. The objective was to determine the feasibility of contracting for road maintenance and to estimate differences in cost, quality, and responsiveness between a contractor and the NSW transportation agency workforce. The resultant findings led the agency to tender for bidders for a 10-year, $130 million performance-based maintenance contract for 450 kilometers of the city of Sydney's roads. Since then, other contracts have been organized similarly.

14.3.5 Learning from Case Studies

These case studies were analyzed to determine how PBC provided value-for-money outcomes for the customer by a series of system elements (Table 14.1).

TABLE 14.1

Elements in a PBC Solution

Element	Detail
Performance measures	The performance parameters (e.g., operational availability) should be clearly identified and measured. The penalties and incentives are defined.
System design of the support solution	The support solution should be designed using systems engineering principles to ensure the performance parameters can be maintained.
Constraints	The constraints to the support solution in meeting the performance measures should be mitigated.
Health measures	Health measures referred to system of continuous monitoring and data acquisition during the contract.
Common threads	Common threads that influence or affect the value for money outcomes must be identified and managed.
Achievement against outcomes	Process of monitoring achievement against outcomes is established.
Risks	The risks of not achieving the performance measures and what is the impact should be identified throughout.
Continuous improvement	Established mechanism to continuously look for opportunities of improvements that provide value for money.

It should be noted that this research has been conducted in Australia. Although the principles are universally applicable, the application context of this research should be adjusted when these elements are applied to specific cases elsewhere.

14.4 Recommendations

From the results of the analysis of the case studies, we have drawn up a series of recommendations.

14.4.1 Recommendation One

A PBC contract not only requires that performance targets are met, but also strives to improve efficiency. Thus, the contractor has to design a service system to fulfill the contract, and in addition must propose an efficiency implementation plan. The following six-step process is recommended for developing and designing a PBC contract from first principles:

1. Specify the outcomes
2. Select the performance measures
3. Set the contracted levels
4. Define the payment regime
5. Define the incentive regime
6. Establish the contractual framework

14.4.2 Recommendation Two

Instead of using a whole-of-system support contract, use a set of smaller support contracts that, in combination, provide the entire support for the system. Additionally, particular items fitted to a system can be the subject of standalone support contracts if they require specialist skills or to make use of support arrangements for items common to multiple systems (e.g., all aircraft tires could be the subject of a single support contract). These approaches represent better value for money overall in a number of ways:

- Enabling competitive tension to be maximized by facilitating competition for the support of lower-level elements of a system.
- Removing elements from higher-level system support contracts where competition for support is restricted.

- Reducing costs by tapping in to existing lower-cost and more responsive supply chains.
- Reducing risk loading (and therefore, costs) because a whole-of-system support contractor does not have to undertake or outsource the support of system elements for which it has little or no experience.

We recommend that a PBC process be developed to provide guidance on how to improve value for money. It should include ways to achieve these objectives:

- Tailor for different PBC outcomes
- Pass down PBC requirements to subcontractors effectively
- Apply risk assessment to the PBC
- Develop a PBC pain/gain-sharing regime

14.4.3 Recommendation Three

Our analysis suggested that for a PBC contract to be cost effective, the contract length needs to be at least five years. This gives the contractor the incentive to pursue engineering innovations that will pay back within the contract period, and could result in cost reductions in future contract extensions. Inclusion of appropriate contract clauses that specify conditions for awarding contract extensions is an effective way of providing incentives for a contractor to incorporate innovations at the earliest possible time.

We recommend developing a PBC policy for the negotiation of initial PBC contract lengths and subsequent extensions.

14.4.4 Recommendation Four

Meeting customer expectations is crucial to the achievement of PBC targets. Since meeting performance outcomes is the primary means of satisfying the customer, it is key to ensure that due diligence is given to the trade-offs necessary to achieve value-for-money performance outcomes in PBC contract design.

In a support environment, performance measures must be very carefully designed. There is always the risk of unintended consequences when the push to achieve a performance measure includes heavy penalties or strong incentives, as this can drive negative behaviors.

We recommend developing a PBC with these objectives:

- Develop and specify outcomes that will provide value for money
- Conduct trade-off analyses that consider value for money

- Match incentives to the requirements
- Decide when the total cost of ownership exceeds the benefit it provides
- Select appropriate performance measures
- Select appropriate performance indicators

14.4.5 Recommendation Five

A typical procurement cycle consists of five stages:

1. Establishing the need
2. Defining the requirements
3. Acquisition of the system
4. Sustaining the system
5. Disposal

It is important to establish the need first in order to build the framework required for the other aspects of the PBC.

The framework for performance-based incentives and the subsequent performance-based payment regime cannot be set until a baseline for performance has been established. In defense contracts, by way of example, this occurs during a transition period which normally occurs no later than 12 months after acceptance of the system.

A transition period has a number of potential benefits:

- Reducing uncertainty about performance targets
- Avoiding performance discrepancies in initial phases when the project stakeholders are not familiar with the system and some abnormal outcomes could be observed.
- Properly gauging equipment reliability

We recommend developing a corporate-level procedural instruction work instruction that can be used as guidance to the establishment of PBC or PPBC baselines. The following are some suggested topics that should be covered:

- Define the level of authority to handle the contract
- Guidance on what to do during the transition period between acceptance of the product and commencement of the penalty period
- How to design the solution to meet the payment and incentive regimes
- Guidance on how to gauge equipment reliability

14.5 Performance Modeling

It is inevitable that a system of systems will need to operate in a PBC regime in the future. A typical four-band performance incentive/penalty scheme with the capability distribution overlaid on the achieved performance axis is shown in Figure 14.1. The capability distribution (the dotted curve) represents the probability density function of the enterprise achieving a performance level on the x-axis. The probabilities of achieving certain performance level in a PBC can then be identified as the area in one of the performance bands as shown in Figure 14.1.

The position of the probability density function relative to the contract performance is therefore an indication of the capability of the company to deliver the expected performance.

Several decisions can be made using this assessment outcome of the relative position of the probability density function. Firstly, based on this information, the prime contractor can decide whether to go ahead with their existing enterprise capabilities. This is a go/no-go decision scenario. The contractor will have to decide in conjunction with other concurrent opportunities, which may be assessed by the same systems-of-systems opportunity assessment methodology, or by other means.

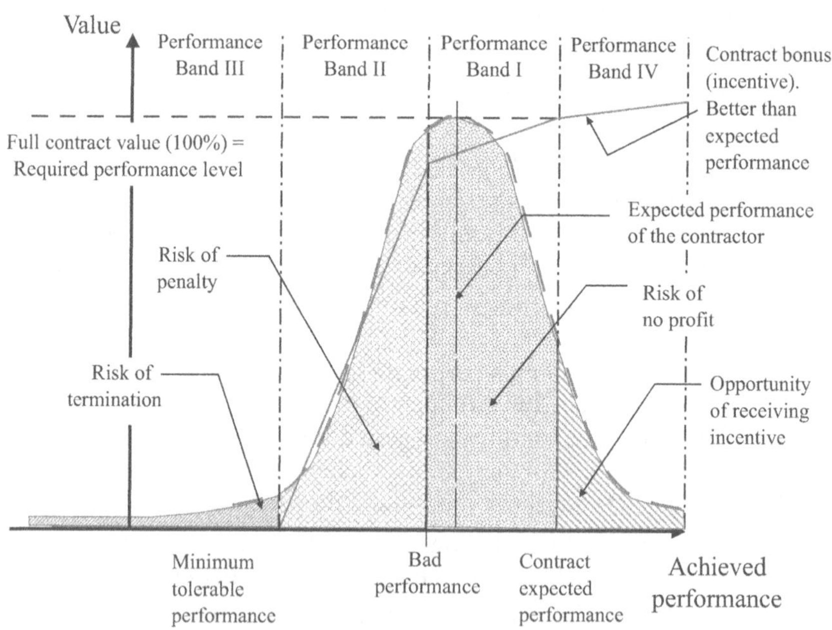

FIGURE 14.1
Performance bands and risks in PBC

Secondly, if the risk level is too high, the prime contractor can acquire additional enterprise capabilities, thereby increasing the cost of providing the system. Inevitably this means raising the contract price to cover the costs, or by implementing organizational improvements such as lean production and Six Sigma. In the latter case, the time factor of the 3PE will be brought in to map the change of capability over time.

Thirdly, the prime contractor can identify the shortfalls in capability and collaborate with other prime suppliers in the industry. The performance capability assessment is then modified to show the overall performance capability of the collaborative network. A mapping of potential changes over time in capabilities of other parties should also be considered, as highlighted by the 3PE model.

Finally, the prime contractor can consider boosting its core capabilities by mergers and acquisition with other companies. This solution is more complicated, since the decision on which companies to acquire depends on strategic alignment requirements. However, this option represents an immediate shift of the capability distribution to the right. The only concern is whether the new organization can be restructured to operate effectively in time to execute the PBC.

The capability model in Figure 14.2 is based on the categorization of capabilities into 3PE elements and interactions. Through a simple modeling process involving both cost and availability, the performance of a service system can be estimated. This model is relevant to the analysis of risk in complex engineering system upgrade projects because it can help to determine where a company should increase capacity, effort, and expenditure to reduce or mitigate risk.

The ability for achieving performance needs to be supported by the ability of the organization to perform. Measurement of capability is very situation specific and difficult to manage. The setting of realistic performance targets in the initial planning phase ensures an acceptably low probability of failure. A realistic plan reduces need for intensive interventions in the middle of the project in order to achieve over-optimistic initial performance targets.

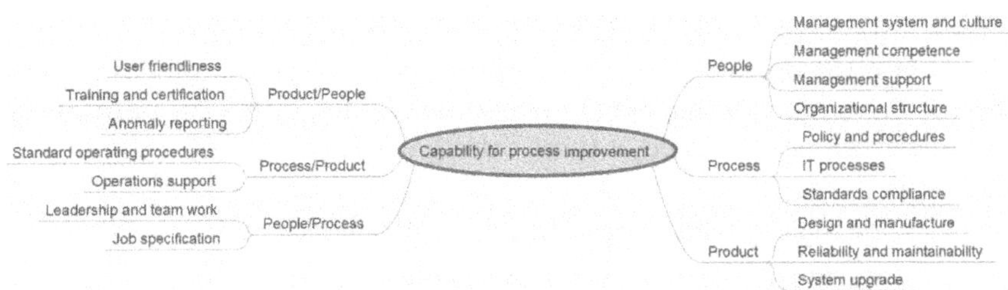

FIGURE 14.2
Process improvement capability score as a function of 3PE architecture elements

To effectively calculate the capability score of a system of systems, it is necessary to break down and categorize the individual capability factors in the people, process, and product elements in the 3PE model so that more accurate assessment of each capability factor can be made. To do this, an enterprise architecture must be chosen to form the modeling framework. In chapter 4, we introduced the 3PE model for enterprise modeling. This is based on the key elements of people, process, and product, and three of their primary interactions, viz, people vs. process, product vs. people, and process vs. product. Changes in the environment in which the enterprise operates drive changes in the internal architecture.

However, there are many parameters that characterize the capability elements. For example, human beliefs, fears, and emotions that affect the degree to which performance can be improved fall into the people category. The people–process interface focuses on processes with a considerable human factor, such as training in new processes and process improvement techniques together with methods for communication of change employed for groups inside and outside the organization. The people–product interface includes capability factors and processes affecting product R&D. The process–product interaction concerns the R&D activities and outcomes of production and product. It is necessary to elaborate the capability indicators at a finer granularity, as shown in Figure 14.2, when the focus area is decided.

This capability is assessed on the scenario of KPIs specific to the relevant projects. If $\{\mathbf{X}\} = \{X_1, X_2, \ldots X_n\}$ is the set of parameters subject to performance measure assessment, and the relative importance of each of the parameters in the set is given by $\{\boldsymbol{w}\} = \{w_1, w_2, \ldots w_n\}$, then

$$C_i = \sum X_i w_i \tag{14.1}$$

$$\sum_i w_i = 1 \tag{14.2}$$

Each parameter will be represented by a normal distribution of the current capability of the company. These distributions are subject to variations, due to the changing environment E, in which the system operates.

To obtain the overall distribution of the system, let's examine the assessment of performance according to known capability levels. The performance outcome X can be estimated at three different levels:

- x_o = the most *optimistic* performance outcome due to x
- x_m = the most *likely* performance outcome due to x
- x_p = the most *pessimistic* performance outcome due to x

The probable performance outcome of a capability factor can then be approximated as a normal distribution curve with these parameters:

$$\text{Mean value: } x_e = \frac{x_o + 6x_m + x_p}{6} \tag{14.3}$$

$$\text{Variance: } v_e = \left(\frac{x_p - x_e}{6}\right)^2 \tag{14.4}$$

$$\text{Standard deviation: } \sigma_e = \frac{x_p - x_e}{6} \tag{14.5}$$

If there are multiple capability factors, the importance of individual factor can be assigned a weight, which will integrate the effects of each factor into overall consideration of the system:

$$\text{Weighted mean: } x_e = \sum_{i=1}^{n} c_i x_i \tag{14.6}$$

$$\text{Weighted variance: } v_e = \sum_{i=1}^{n} c_i \left(\frac{x_{pi} - x_{oi}}{6}\right)^2 \tag{14.7}$$

14.6 A Worked Example

The above methodology can be illustrated with an example. Note that the values are hypothetical and can be adjusted to more realistic values in an actual case. Not all the enterprise elements in the 3PE model have been taken into consideration. In this example, let's assume the targeted performance P of an asset management and support service system is the achievement of two independent performance indicators:

$$\{\boldsymbol{P}\} = \{\text{availability} = 96\%,\ \text{cost} = \$210{,}000\} \tag{14.8}$$

Hypothetical values used for the system are as follows:

- The component affecting the system's availability has mean time to failure (MTTF) T_o of 1000 days.
- The time to repair T_r varies between 1 and 5 days with an average time of 2.5 days.

TABLE 14.2

System Availability due to Variations in Time to Repair

Scenario	Time to Repair (days)	Availability
Best case	1	0.999001
Medium case	2.5	0.997506
Worst case	5	0.995025

- Cost of each component is $10,000. The components are replaced during the annual maintenance service.
- There are four replacements per year (every 90 days) so the component cost is $40,000 per year.
- Total labor cost associated with the service of this system is $80,000 per year.
- Organization overhead is 100% of labor cost.

The availability performance indicator of the system can then be computed by Eqn. 14.9:

$$A = \frac{T_o}{T_r + T_o} \tag{14.9}$$

The availabilities for the best, medium, and worst case scenarios are as shown in Table 14.2.

However, the availability values as shown in Table 14.2 are ideal, and assume that every other part of the system is operating perfectly. Suppose there are some problems with the spare parts supply chain (i.e., a product-process interaction) such that the spare parts are out of stock occasionally. Assuming that, on average, 20 out of every 1000 days are lost waiting for spare parts, the adjusted availability in the three scenarios is then adjusted to the values as shown in the middle column of Table 14.3. Furthermore, it is assumed that management inefficiency (i.e., a people–process interaction) such as documentation delay is estimated to lose 50 out of every 1000 days. The effect on availability is shown in the right-hand column of Table 14.3.

TABLE 14.3

System Availability after Considering Management and Supply Chain Inefficiencies

Scenario	Availability after Considering Spare Part Supply Chain	Availability after Considering Management Inefficiency
Best case	0.979021	0.949051
Medium case	0.977556	0.947631
Worst case	0.975124	0.945274

TABLE 14.4

System Availability after Considering Uncertainties

Performance Outcome	Availability
Optimistic	0.979021
Medium	0.962088
Pessimistic	0.945274

Hence the approximated normal distribution of system availability due to a combination of organizational capabilities is given in Table 14.4. The normal distribution has the following parameters, calculated by Eqns. 14.3 to 14.5:

- Mean = 0.962108
- Variance = 0.0000316
- Standard deviation = 0.005625

For the desired availability performance indicator of 96%, the normalized Z variable is −0.3747 and the probability of meeting the performance is 64.6%. In other words, the probability that this improvement initiative may fail is 35.4%.

This is not a great outcome, but does illustrate the fact that improvement projects can be quite risky.

Now, let's look at the second performance indicator, cost. If the overhead rates are assumed to be different in each of the three scenarios, the cost factors are calculated as shown in Table 14.5.

The cost can be affected by the morale of the workforce (i.e., the people factor). Assume that the labor cost can be reduced by 5% if the morale of the service team is high. On the other hand, if the morale of the service team is low, the labor cost is estimated to increase by 10%. The effect is shown in Table 14.6.

Again, the approximated normal distribution of cost performance is computed as shown in Table 14.7. It should be noted that for the cost performance indicator, success is shown by lower values (as compared to the availability indicator, where a higher value is expected).

TABLE 14.5

Cost of System due to Different Overhead Rates

Scenario	Overhead Rate Change	Cost
Best case	10% reduction	$192,000
Medium case	0% (no change)	$200,000
Worst case	15% increase	$212,000

TABLE 14.6

Cost of System due to Workforce Morale

Scenario	Cost for High Morale (5% labor cost reduction)	Cost for Low Morale (10% labor cost increase)
Best case	$184,400	$207,200
Medium case	$192,000	$216,000
Worst case	$203,400	$229,200

For the desired cost performance indicator of $210,000, the normalized Z variable is 0.732143 and the probability of meeting the performance is 75.8%. This is a relatively good outcome, indicating that people factor and its linked elements are under good control.

Lastly, the weighting of the two performance indicators can be determined by the Delphi method or the analytic hierarchy process. Either weighting method will provide a balanced view of the aggregated performance value. Let's assume that the weighting is determined as follows:

$$\{W\} = \{\text{availability} = 0.6,\ \text{cost} = 0.4\} \tag{14.10}$$

Therefore, the combined performance distribution has an expected mean of 0.69482. The standard deviation of the normalized distribution is 1.0 so the risk that the service improvement project will not meet the expected target performance indicators is assessed as 24.4% (i.e., the probability of success is 75.6%).

The decision-maker of the company must decide whether the risk is acceptable. If the risk is not considered to be acceptable, further investigation can be carried out to explore what the company can do to reduce the risk. For example, by improving the process (e.g., by process mapping) or spare parts stock control (e.g., by eliminating stock-outs).

TABLE 14.7

Normalized Cost Performance Factor Distribution

Performance Outcome	Cost		
Optimistic	$184,400	Mean	$204,533
Medium	$203,400	Variance	$55,751,111
Pessimistic	$229,200	Standard deviation	$7467

14.7 Reflections

This chapter discusses the performance assessment of systems and its implications for systems of systems. As the business world increasingly favors the use of performance-based contracting method to manage complex engineering projects, it is important to understand the effect of performance measures as the basis for business pricing and contractual terms.

Many decisions in life are made on the basis of the performance outcomes of the system under consideration or being used. For example, when you purchase a microwave oven, the user manual will suggest processing times based on the amount and type of materials used. The performance of the microwave oven can be tested by using the suggested settings and observing if your cup of coffee is heated to an acceptable temperature. Similarly, you can find this type of performance indicator for other systems.

Recall some of your experience of measuring the success of a team by its performance outcome. Starting from the most general performance statement, break down your performance indicators as shown in Figure 14.2. Present your results to your good friends and ask them if they think there is anything missing.

14.8 References

1. Bryne T., & Mo, J.P.T. (2015). "Critical factors for design of successful performance-based contracting environment." *International Journal of Agile Systems and Management*, 8(3/4), 305–331.
2. BAE Systems. (2012). Over-The-Horizon Radar, Case Study, Available from http://www.baesystems.com/our-company-rzz/our-businesses/bae-systems-australia/defence-solutions/integrated-solutions/over-the-horizon-radar-case-study/

14.9 Additional Reading

Mo, J.P.T., Bil, C., & Sinha, A. *Engineering Systems Acquisition and Support*. Elsevier, 2015, Chapter 11. ISBN: 978-0-85709-212-0.

15

Managing System Models

> It was the ancient opinion of not a few, in the earliest ages of philosophy, that the fixed stars stood immoveable in the highest parts of the world; that, under the fixed stars the planets were carried about the sun; that the earth, us one of the planets, described an annual course about the sun, . . . However, it was agreed on both sides that the motions of the celestial bodies were performed in spaces altogether free and void of resistance . . . But our purpose is only to trace out the quantity and properties of this force from the phenomena, and to apply what we discover in some simple cases as principles, by which, in a mathematical way, we may estimate the effects thereof in more involved cases: for it would be endless and impossible to bring every particular to direct and immediate observation.
>
> **Isaac Newton**
> *The System of the World, between 1685 and 1727*

15.1 Fundamental Concept of System Modeling

A system modeling tool is a standardized method of describing system models and sub-models with the following features: clarity of representation, variable viewpoints, modular and incremental. In some cases, these tools are also considered to be modeling languages because they specify the forms and relationships by which the system functions and concepts are represented. The modeling languages created through substantial researches in the last half century allow as much flexibility in modeling description as possible while maintaining minimal restriction to the way the forms are put together. Most modeling languages use a building-block approach and support the definition of partial models (e.g., from libraries of partly assembled building blocks).

15.2 Functional Analysis

An engineering system developed using the systems engineering process (see Section 1.4) will eventually become the in-service system being operated by the user. Initial system development requires that all user requirements be identified and system specifications be drawn up to meet these user requirements. From the system specifications, a set of high level functions can be identified and created (sometimes with innovative ideas from research projects elsewhere). The high level functional descriptions are simple sentences that describe what the function will do to depict a visible behavior in the system.

To develop the real components and system, it is important to analyze these functions and design the hardware and software bit by bit, with the objective that these bits and pieces can be tied together later to produce the desired system. This process is the rule of "divide and conquer," i.e., divide the complex system into simpler units so that the simpler units can be conquered (solutions created) easily.

Functional modeling in systems engineering is a structured representation of functions, i.e., activities, actions, processes, and operations, within the modeled system. A function model is a graphical representation of a system's function within a defined scope. The purpose of the function model

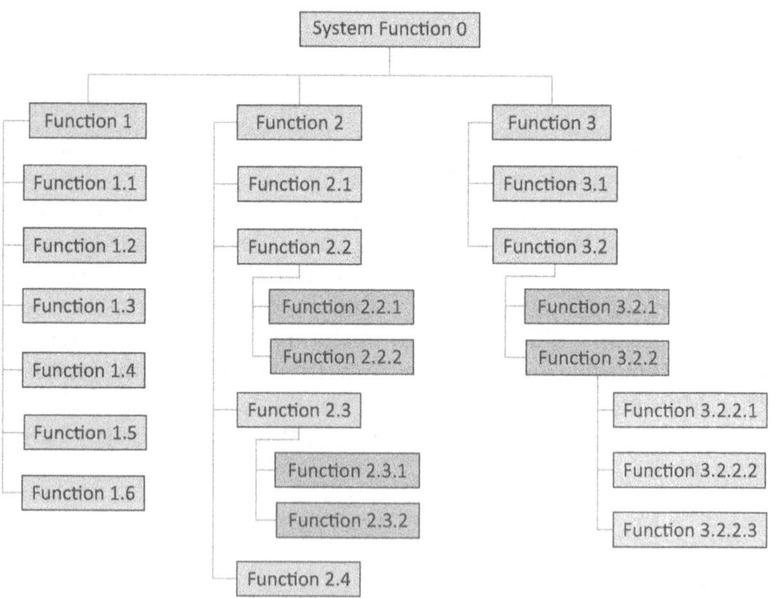

FIGURE 15.1
Hierarchy of functions

is to describe the functions and processes, assist with discovery of information needs, help identify opportunities, and establish a basis for determining product and service costs.

Functional decomposition refers broadly to the process of breaking down a functional relationship into its constituent parts in such a way that the higher level function can be understood as a series of connected, simpler functions. This decomposition process is undertaken either for the purpose of gaining insight into the constituent components or to achieve a certain level of modularity. The basic idea is to try to divide a system, starting from the top "system level," in such a way that each block of the block diagram can be described as a linked set of functions inheriting the same input, control, mechanism, and output of the higher level function. This top down modeling approach enables the complexity of a system to be unwrapped through a number of simple steps. At the start of the modeling work, we represent the whole system as one block. Next, we go into details to decompose the block into several linked sub-blocks. Each of the sub-blocks is further decomposed to the next level (Figure 15.1).

The decomposition process will divide the top function into many levels of sub-functions. The hierarchy of functions can be represented by a tree structure as shown in Figure 15.1. By the same token, the functions are labeled in hierarchical numbering sequence accordingly. Functions at the same level are indicated by a single color.

15.3 Functional Flow Block Diagrams

A functional flow block diagram (FFBD) is one of the primary analysis tools for understanding how functions work as a group (Figure 15.2). The purpose is to show the sequential relationship of all functions that must be accomplished by a system. Each function (represented by a block) is identified and described in terms of inputs, outputs, and interfaces from the top down so that sub-functions within the function are recognized as a part of the larger functional area. Within the boundary of the function, sub-functions may be performed in parallel, or alternate paths may be taken. Functions are arranged in a logical sequence so that any specified operational use of the system can be traced in an end-to-end path. The FFBD network shows the logical sequence of "what" must happen, but does not assume a particular answer for "how" a function will be performed.

To illustrate a FFBD, we need to start with the function hierarchy. Figure 15.2 shows the function hierarchy of a zone defense system with three main functions: (1) Detect threat; (2) Eliminate threat; (3) Re-evaluate threat.

From the hierarchy, the "Detect threat" function has ten sub-functions, each performing some specialized activity. Similarly, the "Eliminate threat"

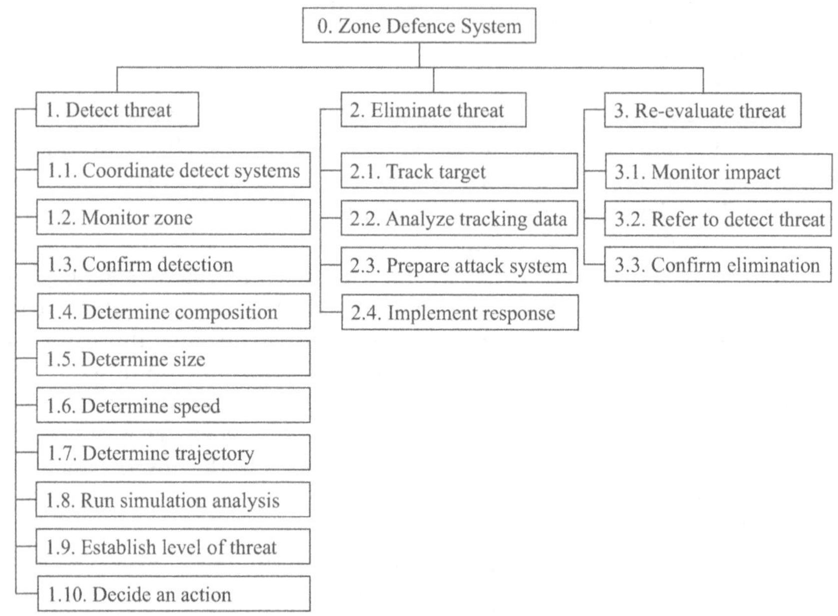

FIGURE 15.2
Zone defense system hierarchy of function

function has a weapon system that can eliminate the threat if activated. The "Re-evaluate threat" function is a post-mortem function that checks whether the threat has been eliminated. Sub-function "Refer to detect threat" directs the system to certain sub-functions of the "Detect threat" function because there is no need to duplicate parts of the system that can already do the job.

The hierarchy of function diagram does not show how and when the activities are carried out. The FFBD shows additional information as shown in Figure 15.3.

From Figure 15.3, it is now clear that sub-functions 1.4, 1.5, 1.6, and 1.7 can be performed in parallel. The AND junction before the parallel sub-functions indicates that the confirmed detection data are passed to all four parallel sub-functions, each of which carries out a particular type of analysis. The AND junction after the parallel sub-functions indicates that the analyzed information is required to feed simultaneously into a simulation analysis. The result of the analysis is used to establish the level of threat. Function 1.10 determines if the decision is to eliminate the threat, in which case the attack system is activated (external function 2). If the decision is to wait (i.e., the threat is not serious), the system returns to monitoring mode by activating sub-function "Monitor zone." The OR junction indicates the effect of this decision.

This example shows that the FFBD originates from the hierarchy of functions. It contains operational logic and indicates when (in terms of work flow) certain functions or sub-functions are in action.

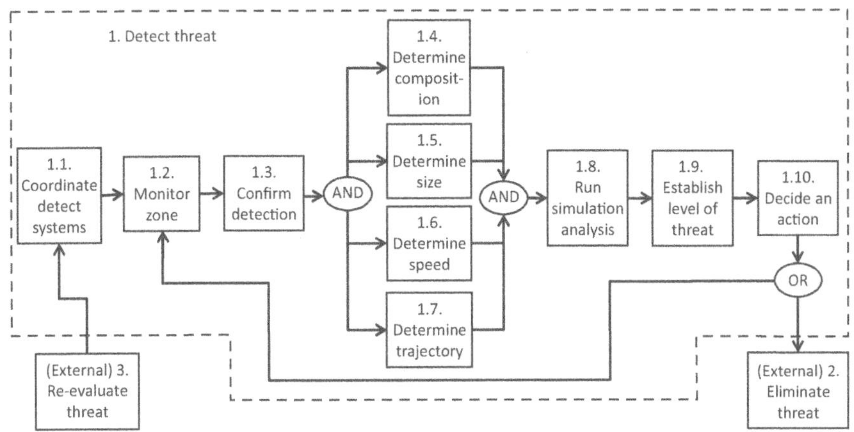

FIGURE 15.3
Functional flow block diagram of a zone defense system

15.4 Data Flow Diagrams

A data flow diagram (DFD) shows the types of data, the data path, actions taken based on the data, and where the data is stored. Data flow diagrams are an integral part of the hierarchical functional modeling process. When a function is decomposed to its lower levels, the corresponding data flows in and out of the function are also decomposed to represent data flowing in and out of the lower-level functions.

A DFD consists of four components:

- Process: Illustrates the transformation from input to output.
- Store: Represents a data collection or some sort of material
- Flow: Shows the movement of data or material in the process
- External entity: An entity that interacts with the modeled system, but is external to it.

The DFD in Figure 15.4 uses the hierarchy of functions of the zone defense system in Figure 15.3 as its basis. In some respect, it looks similar to the FFBD but the main point of the DFD is to indicate data flow between the functions.

Starting with the sub-function 1.1, this can either be activated automatically or by an external command from function 3. The "Coordinate detect systems" sub-function issues a monitor command to sub-function 1.2. Another source of the data is the action decided, and hence the information is ORed to the sub-function "Monitor zone." Data from sub-function 1.3 is directed to four

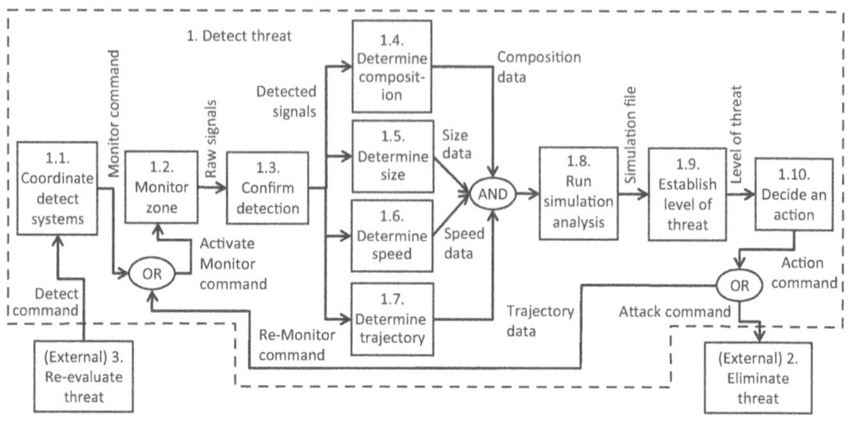

FIGURE 15.4
Data flow diagram of a zone defense system

parallel sub-functions simultaneously. Each of the parallel sub-functions produce different data which are grouped together (ANDed) enabling the simulation analysis to run. The rest of the data flow is self-explanatory.

15.5 Petri Net

Petri net is a theory of systems organization created by Carl Adam Petri. It has a strong emphasis on set theory forming the basis for mathematical analysis. Petri net is a network model that has been used extensively by system designers and analysts to identify key processes within discrete event-processing environments. It consists of two key elements:

- Net Structure: A weighted-bipartite directed graph that represents the static part of the system.
- Marking: Represents a distributed overall state of the structure.

Formally, a Petri net can be defined as $N = (P, T, I, O, M)$ where

P = a finite set of places

T = a finite set of transitions

I = an input function such that $I: P \times T\ N$

O = an output function such that $O: T \times P\ N$

M = a marking vector with the same dimension as P indicating the number of tokens at places in the network

Places are used to represent availability of resources, operation processes, or conditions. They are depicted as circles. Transitions are used to model events. In an automated system, an event can be start or termination of an operation and are depicted as bars.

To model a discrete-event dynamic system, we need to take into account its states and the events leading to evolution of states of the system. In a Petri net, the state is described by means of a set of state variables representing local conditions. Hence, states are associated with places and transitions which are the entities that can effect changes. Places and transitions are related through a weighted flow relation indicated by the arrows in the network. The flow relation is modeled by arrows joining related places and transitions. A weight known as a "token" can be assigned to the arc to indicate the numerical importance of that relationship.

Let's look at a Petri net example.

Consider the maintenance process of an aircraft in the hanger. To start the process, an aircraft is scheduled to be available for maintenance. When the hanger is vacant, the aircraft is moved into the hanger. At the hanger, service and maintenance operations are performed on the aircraft. A signal is passed to the hanger operator when all maintenance works are complete. The aircraft is then moved out of the hanger, and is then made available for use. The hanger returns to the state of being available. This process is modeled by four places and two transitions in Figure 15.5.

Petri nets can also be hierarchical. For example, if we need more details of the maintenance operation in p_2, the place can be expanded further as shown in Figure 15.6. In this case, the maintenance required is the repair the aircraft engine. The cover has to be disassembled in p_{21} (note: in older aircraft, this can be much more complicated than just unlocking

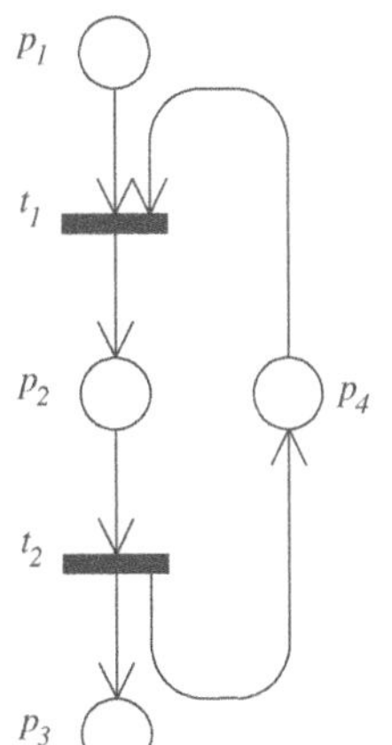

4 places:
p_1 = aircraft available for service
p_2 = aircraft maintenance operation being performed
p_3 = aircraft maintenance complete and available for use
p_4 = hanger available

2 transitions:
t_1 = aircraft moved into hanger
t_2 = aircraft moved out of hanger

FIGURE 15.5
An example of a Petri net

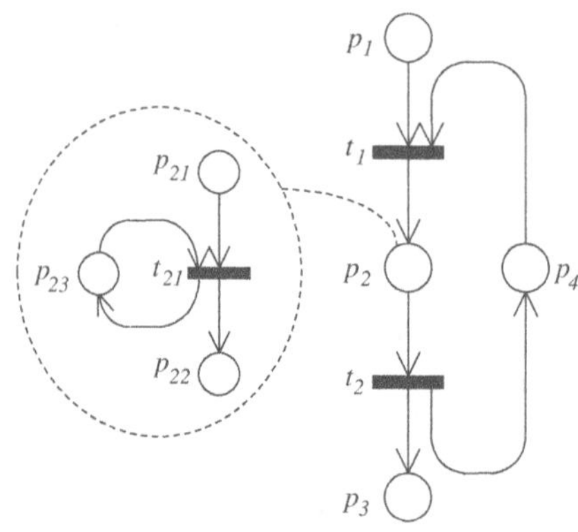

4 places:
p_1 = aircraft available for service
p_2 = aircraft maintenance operation being performed
p_3 = aircraft maintenance complete and available for use
p_4 = hanger available
p_{21} = disassemble aircraft engine cover
p_{22} = assemble aircraft engine cover
p_{23} = repair engine

2 transitions:
t_1 = aircraft moved into hanger
t_2 = aircraft moved out of hanger
t_{21} = engine accessible

FIGURE 15.6
An example of a Petri sub-net

or unscrewing a cover plate). When the engine is exposed, the engine repair can be performed p_{23}. When the repair is complete, the cover can be re-assembled and the process is returned to the main Petri net.

15.6 Unified Modeling Language

Unified Modeling Language (UML) is an industry standard of the software industry. Three leading object-oriented methods were used prior to UML:

- Booch and Gray—originator of object
- OMT—Object modeling techniques
- OOSE—Object-oriented software engineering

Each of these methods has its merits and shortcomings. In order to develop a more comprehensive methodology, the Object Management Group standards consortium was formed to unify the three methods into UML. Useful constructs in UML include sequence diagrams, state charts, and class diagrams. More information on UML can be found at http://www.uml.org/.

UML has many modeling tools that can be used to represent different views of a system. We use a system in a telecommunication company as an example to illustrate the methodology. The use case diagram of the message system is shown in Figure 15.7.

Like other function-modeling tools, UML can be used to represent a telecom system by a hierarchy of functions as shown in Figure 15.8.

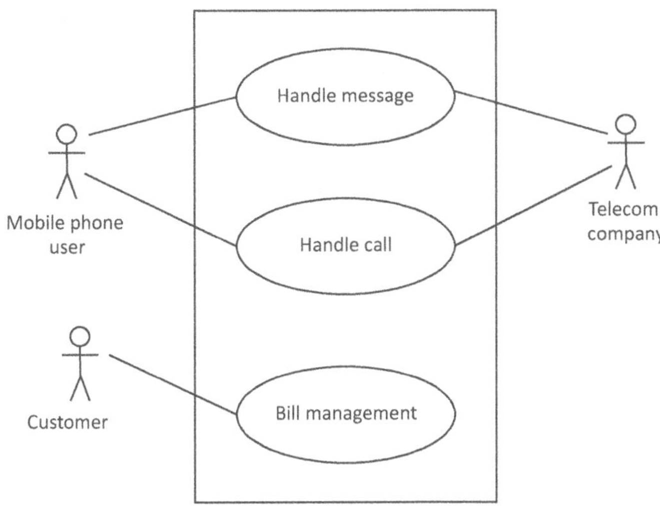

FIGURE 15.7
Use case of a telecom system

Now, we need to use UML to expand into details. Let's use the function "Handle call." This can be developed into a number of classes generally structured around a generic class known as "Call management," as shown in Figure 15.9. A class has attributes and methods, which are entered into the class box. Attributes are variables used by the methods. In the class "Handle call," the attributes can be the caller's name, time of call, etc. Methods are routines that perform the function of the class.

In Figure 15.9, the "Call management" class contains a set of attributes and methods that will be used to handle calls. This class is then instantiated (copied into the program) and enhanced with specific attributes and methods. In this case, the "Initiate call" class is instantiated from the "Call management" class and is further enhanced to handle information when a call is initiated. For example, the "Initiate call" class will find out the caller's information such as phone number, location (for wireless connection), number to call, etc. This information would probably reside somewhere in

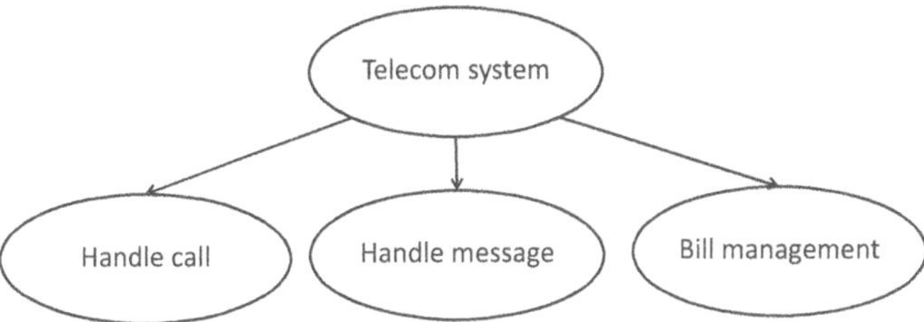

FIGURE 15.8
Hierarchy of functions represented in UML format

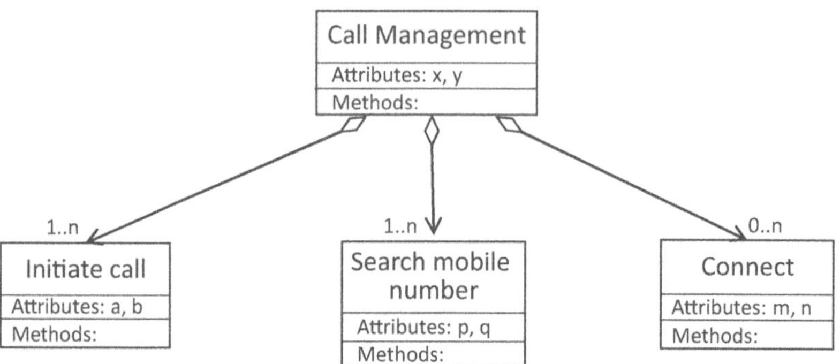

FIGURE 15.9
Class diagram for "Call management"

the communication messages. The arrow (with a diamond at the other end) shows the parent ("Call management") and child ("Initiate call") relationship. It is also annotated with 1..n which means we can have more than one instance of "Initiate call" within the memory of the system (because we may have many actors, i.e., users, initiating calls simultaneously).

The class "Search mobile number" will search for the route that can be used to connect to the number to be called. This can be a wide-area search,

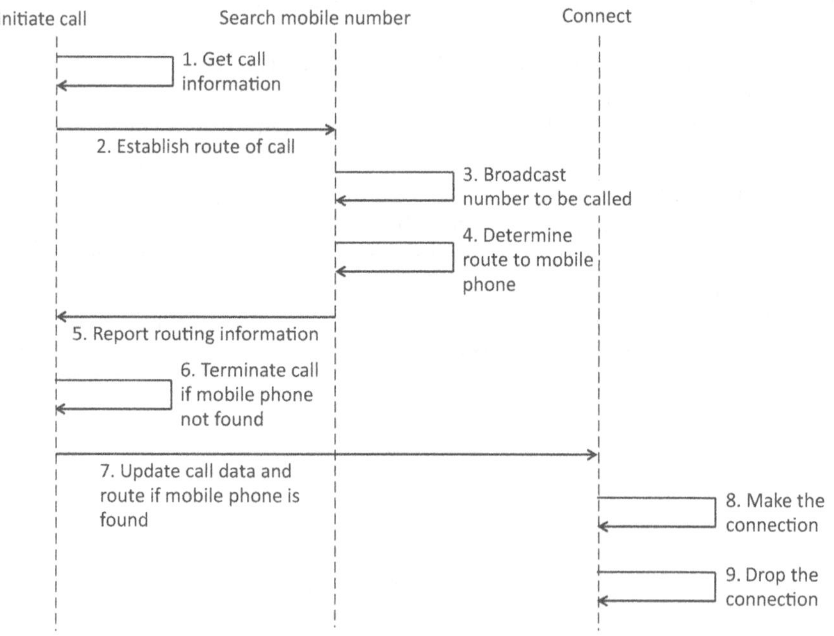

FIGURE 15.10
Sequence diagram of the "Handle call" class

possibly international. Again, multiple instances of this search are required to handle different calls. If the mobile phone number cannot be found, the routing information will indicate a non-reachable phone and the class "Initiate call" will return a message notifying the caller that the mobile number is not in range. If the routing information is determined, the class "Connect" will be started to make the connection. The number of instances for class "Connect" can vary from zero (no connection) to many connections.

In UML, function flow is represented by a sequence diagram. A sequence diagram indicates the sequence of activities between the classes. Figure 15.10 shows the sequence of activities to handle a call from one mobile phone to another.

The activities are indicated by the arrows and labeled with the necessary actions to be taken. The convention is to start from top of the first activity (from any class) and sequentially work through the classes in the diagram as shown by the arrows. The activities are numbered to indicate the order of execution.

15.7 Reflections

The system modeling tools outlined in this chapter are best used at the initial design and planning stages of developing a system of systems. The functional diagrams and flow-diagram tools use traditional functional modeling concepts, and hence are easily applied. The UML modeling system is derived from software engineering projects but its generic nature allows users to apply the methodology to the development of systems of systems.

Select a system of systems in your working environment. Use both the functional diagram method and UML to model the same system of systems. Compare the differences between the two methods and highlight their pros and cons.

15.8 Additional Reading

Cook, M. & Mo, J.P.T. (2015). "Strategic risk analysis of complex engineering system upgrades." *European Scientific Journal, October 2015, SPECIAL edition*, 2, 64–80.

Nicholds, B. & Mo, J.P.T. (2013). "Determining an action plan for manufacturing system improvement: The theory." *International Journal of Agile Systems and Management*, 6(4), 324–344.

Nicholds, B., Mo, J.P.T., & Bridger, S. (2014). "Determining an action plan for manufacturing system improvement: A case study." *International Journal of Agile Systems and Management*, 7(1), 1–25.

Index

T

U

V

W

Z

Milton Keynes UK
Ingram Content Group UK Ltd.
UKHW040444071024
449327UK00020B/977